Animal Biology: Taxonomy, Anatomy and Physiology

ABOUT THE AUTHORS

Dr. Kapil Kishor is Professor & Head of the Department of Zoology of Mahila Shilp Kala Bhawan College, Muzaffarpur (Bihar). He has teaching and research experience about 16 years. His 21 research papers have been published in different national and international research journals. He authored books of Zoology, Biochemistry, Biotechnology and Environmental Sciences. Dr Kishor has received 10 Follow-up Awards of Academics bodies like Zoological Society of India, Indian Academy of Environmental Sciences, Society of Life Sciences, International Consortium of Contemporary Biologists, Beloved Research Society, Applied Zoologists Research Association etc. He is also recipient of Young Scientist Award, Senior Scientist Award, Distinguished Scientist Award, Eminent Scientist Award, Jewel of India, Bharat Jyoti Award, Glory of India Gold Medal, Indira Gandhi Award, Best Personalities of India Award etc. He has attended many National and International seminars.

Sofia John is Professor of Microbiology. After Ph.D. in Cellular and Molecular Immunology from Poland, She worked for Ten years in the field of Immunology and Microbiology, Cellular and Molecular Immunology and Plant Science and Zoology department. She also took up a lectureship in Plant Science, and worked in the area of Microbiology and Immunology, agriculture microbial and ecology investigating microbial antagonism and production of antibiotics from bacteria. She has authored 9 books and over 31 articles and research papers and has presented several papers in various National and International seminars. She is presently Assistant Professor in Imperial College London is a public research university located in London.

ABOUT THE BOOK

As a biological science, the animal biology major requires lower division preparation in science, calculus and statistics; additionally, you will take courses in the application of biological principles to animals in natural and agricultural systems. Whether as valuable economic resources, beloved companions, destructive nuisances or members of complex natural systems, animals play important roles in human economy, society and culture. Animal biology, the study of biological principles as they apply specifically to animals, bridges the gap between general biological sciences and applied animal husbandry techniques. As an animal biology major, you'll have the chance to put your theoretical knowledge into practice by working with wild and domesticated animals in hands-on situations. The study of animal behavior begins with understanding how an animal's physiology and anatomy are integrated with its behavior. Both external and internal stimuli prompt behaviors — external information (e.g., threats from other animals, sounds, smells) or weather and internal information (e.g., hunger, fear). The book: The method of presentation adopted has been to emphasize the broad general principles of animal biology with a learning towards the physiological aspect.

ANIMAL BIOLOGY
Taxonomy, Anatomy and Physiology

Dr. Kapil Kishor
Sofia John

WESTBURY

WESTBURY PUBLISHING LTD.

ENGLAND (UNITED KINGDOM)

Animal Biology: Taxonomy, Anatomy and Physiology
Edited by Dr. Kapil Kishor & Sofia John
ISBN: 978-1-913229-22-1 (Hardback)

Published by Westbury Publishing Ltd.
Address: 6-7, St. John Street, Mansfield,
Nottinghamshire, England, NG18 1QH
United Kingdom
Email: - info@westburypublishing.com
Website: - www.westburypublishing.com

Printed and bound in India by Replika Press Pvt. Ltd.

For more information regarding Westbury Publishing Ltd and its products, please visit the publisher's website- www.westburypublishing.com

Preface

Animal biology is the study of the chemical and physical makeup of animals. If you choose to study in the field,

Whether as valuable economic resources, beloved companions, destructive nuisances or members of complex natural systems, animals play important roles in human economy, society and culture. Animal biology, the study of biological principles as they apply specifically to animals, bridges the gap between general biological sciences and applied animal husbandry techniques. As an animal biology major, you'll have the chance to put your theoretical knowledge into practice by working with wild and domesticated animals in hands-on situations.

The field of animal physiology extends the tools and methods of human physiology to non-human animal species. Plant physiology also borrows techniques from both fields. Its scope of subjects is at least as diverse as the tree of life itself. Due to this diversity of subjects, research in animal physiology tends to concentrate on understanding how physiological traits changed throughout the evolutionary history of animals.

Animal behaviour, the concept, broadly considered, referring to everything animals do, including movement and other activities and underlying mental processes. Human fascination with animal behaviour probably extends back millions of years, perhaps even to times before the ancestors of the species became human in the modern sense. Initially, animals were probably observed for practical reasons because early human survival depended on knowledge of animal behaviour. Whether hunting wild game, keeping domesticated animals, or escaping an attacking predator, success required intimate knowledge of an animal's habits.

Hormone-dependent sex differences in growth rate have been known for a long time. It has also been known that growth rate and FCE (feed conversion efficiency) are higher in intact males than in castrates. It was natural, then, that the availability of hormones and other natural or synthetic substances displaying hormonal activity led to experiments aiming at their use to increase production. Beginning in the mid-1950s, DES (diethylstilboestrol) and hexoestrol were administered to cattle increasingly in the US and the UK respectively, either as feed additives or as implants, and other types of substances also gradually became available.

This is a reference book. All the matter is just compiled and edited in nature, taken from the various sources which are in public domain.

The book: The method of presentation adopted has been to emphasize the broad general principles of animal biology with a learning towards the physiological aspect.

—Editor

Contents

1

Animal Behaviour

LEARNING BEHAVIOR

Imprinting describes any kind of phase-sensitive learning (i.e., learning that occurs at a particular age or life stage) during which an animal learns the characteristics of some stimulus, which is therefore said to be "imprinted" onto the subject.

The best known form of imprinting is *filial imprinting*, in which a young animal learns the characteristics of its parent. Lorenz observed that the young of waterfowl such as geese spontaneously followed their mothers from almost the first day after they were hatched.

Lorenz demonstrated how incubator-hatched geese would imprint on the first suitable moving stimulus they saw within what he called a critical period of about 36 hours shortly after hatching. Most famously, the goslings would imprint on Lorenz himself (more specifically, on his wading boots).

Sexual imprinting, which occurs at a later stage of development, is the process by which a young animal learns the characteristics of a desirable mate. For example, male zebra finches appear to prefer mates with the appearance of the female bird that rears them, rather than mates of their own type (Immelmann 1972).

Reverse sexual imprinting has also observed: when two individuals live in close domestic proximity during their early years, both are desensitized to later sexual attraction. This phenomenon, known as the *Westermarck effect*, has probably evolved to suppress inbreeding.

Relation to comparative psychology

In order to summarize the defining features of ethology, it might be helpful to compare classical ethology to early work in comparative psychology, an alternative approach to the study of animal behavior that also emerged in the early 20th century. The rivalry between these two fields stemmed in part from disciplinary politics: ethology, which had developed in Europe, failed to gain a strong foothold in North America, where comparative psychology was dominant.

Broadly speaking, comparative psychology studies general processes, while ethology focuses on adaptive specialization. The two approaches are complementary rather than competitive, but they do lead to different perspectives and sometimes to conflicts of opinion about matters of substance:

- Comparative psychology construes its study as a branch of psychology rather than as an outgrowth of biology. Thus, where comparative psychology sees the study of animal behavior in the context of what is known about human psychology, ethology situates animal behavior in the context of what is known about animal anatomy, physiology, neurobiology, and phylogenetic history.
- Comparative psychologists are interested more in similarities than differences in behavior; they are seeking general laws of behavior, especially relating to development, which can then be applied to all animal species, including humans. Hence, early comparative psychologists concentrated on gaining extensive knowledge of the behavior of a few species, while ethologists were more interested in gaining knowledge of behavior in a wide range of species in order to be able to make principled comparisons across taxonomic groups.
- Comparative psychologists focused primarily on lab experiments involving a handful of species, mainly rats and pigeons, whereas ethologists concentrated on behavior in natural situations.

Since the 1970s, however, animal behavior has become an integrated discipline, with comparative psychologists and ethological animal behaviorists working on similar problems and publishing side by side in the same journals.

Recent developments in the field

In 1970, the English ethologist John H. Crook published an important paper in which he distinguished *comparative ethology* from *social ethology*. He argued that the ethological studies published to date had focused on the former approach—looking at animals as individuals—whereas in the future ethologists would need to concentrate on the social behavior of animal groups.

Since the appearance of E. O. Wilson's seminal book *Sociobiology: The New Synthesis* in 1975, ethology has indeed been much more concerned with the social aspects of behavior, such as phenotypic altruism and cooperation.

Research has also been driven by a more sophisticated version of evolutionary theory associated with Wilson and Richard Dawkins.

Furthermore, a substantial rapprochement with comparative psychology has occurred, so the modern scientific study of behavior offers a more or less seamless spectrum of approaches—from animal cognition to comparative psychology, ethology, and behavioral ecology. *Evolutionary psychology*, an extension of behavioral ecology, looks at commonalities of cognitive processes in humans and other animals as we might expect natural selection to have shaped them. Another promising subfield is *neuroethology*, concerned with how the structure and functioning of the brain controls behavior and makes learning possible.

ANTIPREDATOR BEHAVIOUR

Predation is an important source of mortality in the wild. Many species of mammals and flightless birds have gone extinct on islands as a result of introduced domestic cats, rats, and mongooses. These endemics lost certain aspects of predator avoidance behaviour as they evolved on islands without terrestrial predators; preliminary evidence suggests that some continental prey species are also losing their ability to respond to cues of predators in places where large carnivores have been extirpated.

Understanding antipredator behaviour may allow us to make predictions about the fate of populations subjected to introduced predators or conservation manipulation. Controversial data suggest that black rhinoceros Diceros bicornis calves suffer higher mortality in the presence of large carnivores if their mothers are dehomed as part of a conservation procedure, presumably as a result of mothers' inability to fend off predatory attack. Prior knowledge of maternal antipredator behaviour and the function of horns in females would have spoken against dehorning. Since grouping confers advantages on individuals through early detection of predators, warning conspecifics of danger, reducing vigilance time, and diluting predation risk variation in patterns of grouping make species and age-sex classes differentially vulnerable to hunting by humans.

FitzGibbon discusses how antipredator strategies affect the probability of being selected by subsistence hunters. She notes that individuals respond to predation risk by reducing time spent in other activities such as feeding but the extent to which human hunting pressure indirectly affects reproductive rates in this way remains unexplored. Whether antipredator behaviour is principally innate or is modified by experience is critical to captive release

Programmes and some attempts have been made to train captive-bred individuals to recognize and avoid predators.

Although we know that predator recognition is innate in some species such as motmots *Eumomota superciliosa* but must be learned in others such as vervet monkeys *Cercopithecus aethiops* and Belding's ground squirrels *Spertnophilus beldingi* the ecological circumstances under which different sorts of predator template development occurs are still poorly understood. Innate recognition may facilitate rapid antipredator responses, but it may limit our ability to transfer local stocks between areas with different predators. Learned predator recognition, in contrast, may result in high mortality initially, but it does allow reintroduced prey to respond to predators or free-living prey to cope with exotic or reintroduced predators.

FORAGING BEHAVIOUR

Foraging behaviour affects population persistence in the general sense that carnivores are at greater risk of extinction than herbivores, and that, within a dietary class, specialist foragers are more extinction prone than generalists. Thus knowledge of the food requirements of rare species is useful.

At a general level, habitat preferences contingent on feeding requirements will influence species' responses to habitat disturbance and fragmentation. More specifically, foraging theory has shown how patterns of interference between competitors and depletion of feeding patches predict the distribution of individuals and hence the number that a site can hold. Moreover, models have been developed to predict the distribution of animals in patches of differing profitability when individuals are equally competitive or when some dominate others at resources. In theory, these tools might allow conservation biologists to predict consequences of habitat fragmentation if factors such as food abundance, food distribution, and individuals' competitive abilities were known. Monaghan has shown how patterns of feeding in seabirds such as foraging trip length and number of dives can be used to monitor marine fish populations.

DEMOGRAPHIC VARIABLES

Behavioural ecological field studies usually have records on the age and sex of individuals, so it is relatively easy to construct the age structure of the population and the sex ratios of juvenile and adult segments. Estimates of Ne are greatly influenced by population demographics such as these.Creel uses long-term demographic records from social carnivore species to investigate the demographic factors affecting Ne. Because behavioural ecologists are interested in the consequences of traits on individual reproduction and survival, studies usually strive to collect detailed records on individuals' age at first reproduction, interbirth intervals, litter sizes, offspring sex ratios, and longevity, although information on age-specific mortality is more difficult to collect. Life history variables are necessary for the construction of PVAs used to calculate both demographic and genetic MVPs.

Although influential and early PVAs used data collected by field biologists to make predictions about population persistence they were based on relatively crude empirical data. In the context of politically charged controversies where models are scrutinized closely, resulting recommendations may be called into question.

Detailed information on individual life histories can remedy this because they provide an exact measure of variance, which greatly increases the resolution of PVA and Ne calculations.Durant highlights the importance of life history variables in affecting population persistence using three mammal species with different social structures. Her models show that skewed sex ratios are a powerful indicator of extinction and that extending adult survival is effective in increasing population growth rates.

COMPETITION BETWEEN MALES

Certain behavioural and morphological traits are solely used in competition for access to members of the opposite sex. Contests between males have led to the evolution of weaponry such as beetle and deer antlers, to infanticidal behaviour, in which males kill offspring fathered by competitors, and to mate guarding. In some species, high costs of fighting have selected for polymorphic fighting tactics in which some males compete severely over females, whereas others avoid confrontation and mate surreptitiously. The outcome of contests between males will affect who breeds and hence variance in male reproductive success, which affects Ne. Another example of how variance in male reproductive success affects genetic components of populations comes from bison *Bison bison* in the Badlands National Park. There, one lineage is currently being lost because males of that lineage are more timid and less able to defend females than those of the more successful line. Knowledge of patterns of mating and the cues that individuals use to select mates can also help identify the potential for hybridization between species; because one of the goals of conservation is species integrity and persistence, hybridization is normally viewed with alarm. Greene, Umbanhowar, Mangel, and Caro show that in infanticidal species where males kill offspring fathered by other males, population growth rates are reduced, and it may be difficult for such species to withstand exploitation. In theory, Ne will be lowered in species in which infanticide or mate guarding is prevalent.

Different fighting tactics are usually associated with different mortality schedules; thus the frequency with which males display different behavioural and morphological tactics will affect population growth rate and Ne. Rubenstein draws attention to the fact that those Atlantic salmon *Salmo salar* migrating to the sea are more likely to be caught by fishermen than conspecifics that remain behind and mature in their natal streams and that this differential offtake selects for philopatric phenotypes. In a number of reef-dwelling fishes, intersexual competition for mating opportunities has led to the evolution of protandry, in which individuals change from male to female, or protygyny, in which they change from female to male.

Sex change may be precipitated by reaching a certain age or size or by social circumstances when it pays individuals to become the currently rarer sex. Vincent and Sadovy argue that exploitation of larger fish for the tropical fish trade may remove males of certain species, for example, from a population. Thus the speed with which females are able to change to being male will be crucial in determining a population's response to exploitation.

ASPECTS OF ANIMAL BEHAVIOUR

Territoriality and Spatio-Structural Requirements

Hediger first noted the importance of territoriality and functional space for captive management of wild animals. Knowledge of species-specific movement patterns and territory requirements is essential for maintaining close-to-natural social and reproductive interactions in mammals. In addition, perceived space, including olfactory and auditory components, should be considered together with specific niche requirements for successful maintenance and breeding of captive mammals. Providing appropriate space objects (*e.g.*, climbing structures, hiding places) has been shown to have substantial behavioural effects by reducing aggression and stress in several species.

For example, in a pair of giant pandas *Ailuropoda melanoleuca,* a significant decrease in frequency of agonistic vocalisations occurred after addition of a complex series of platforms, swinging tires, and other mobile features to the enclosure. In leopard cats *Felis bengalensis,* reduced exploratory behaviour was an indicator of chronically elevated adrenocortical activity (indicating stress), and hiding behaviour increased with elevated adrenocortical output.

Consequently, provision of appropriate hiding places stimulated exploratory behaviour and reduced pacing and urinary cortisol levels. These results indicate that structural enrichment of enclosures may increase behavioural opportunities for coping with aversive stimuli. Cage size has been shown to affect behaviour and reproduction when crowding occurs. Groups of rodents and non-human primates in captivity exhibited higher frequencies of aggressive behaviour than less crowded or free-living conspecifics. Blake and Gillett found that an increase in cage size resulted in larger litter sizes and increased overall breeding success in Asian chipmunks Tamias sibiricus.

In other rodents, cannibalism and delayed implantation have been observed as a result of crowding. Even when cage size is appropriate, sounds and odours from adjacent enclosures may lead to a perceived crowding effect or territory invasion reported that in an off-exhibit breeding facility for golden lion tamarins *Leontopithecus rosalia* where an air-handling system inadvertently also moved odours and sounds from enclosure to enclosure, neighbouring family groups frequently seemed agitated and vocalised in a distressed manner. Among primates an inappropriate captive environment is frequently associated with stereotyped movement, self-aggression, higher levels of aggression between cage mates, inactivity, and infertility. Indeed, numerous studies on a variety of mammal species have documented an increase in natural behaviour patterns and a decline is stereotypic movements after providing various enrichment devices and increasing the complexity of the captive environment. In some cases reproduction may be directly and positively affected by such changes. For example, several previously non-reproductive gorillas (*Gorilla gorilla*) became reproductively active after transfer to an enriched enclosure. For some species certain enclosure features may be necessary simply to increase survivorship in captivity.

For example, in several ungulate species fatalities have been reported due to inadequate enclosures or enclosure size. Death resulted from collision with fences or other enclosure structures.

In pronghorn *Antilocapra americana* the importance of distant-visibility has been emphasised, and recommen-dations for fencing and shelter constructions have been made to accommodate natural behaviours and reduce fatalities. Furthermore, species-specific enclosure design needs to consider appropriate hiding places, denning sites, substrate, and other structural features. For example, the giant panda spends little time auto-grooming but requires soil or grass and water to allow for regular rubbing and bathing to maintain a healthy fur condition.

INTERSPECIFIC INTERACTIONS

Some species are maintained in mixed-species exhibits. This practice is particularly common in ungulates. However, excessive aggression and even fatalities have sometimes occurred as a result. Zebras (*Equus* spp.), for example, have been observed to kill wildebeest (*Connochaetes* spp.) and giraffe calves (*Giraffa* spp.) at some facilities.

In a multi-institutional study on interspecific aggression in 10 ungulate species, a correlation between aggressive behaviour and taxonomic relatedness was found. Interestingly, aggression was found to be highest between distantly related species. A possible explanation offered for this phenomenon was the inability of distantly related species to recognise each others' threat displays and thus avoid further aggressive interaction.

However, there are encouraging accounts of various attempts to establish mixed-species exhibits, and some have reported an increase in the behavioural repertoire shown by species under mixed-housing conditions.

For small to medium-sized predators, such as cheetahs *Acinonyx jubatus* and a number of small felid species, it has been reported that housing next to large-sized predators and natural competitors, such as lions *Panthera leo* and leopards *Panthera pardus,* may decrease reproductive success. Although recently there has been increased awareness that some species may not thrive well in captivity if in constant, close proximity to their natural predators there have been few systematic studies to investigate this contention.

One study indicates that ungulates may change their behaviour as a result of constant visual presence of predators. However, the authors concluded that because the changes only occurred in rare behaviours, juxtaposing predators and their prey in exhibits did not result in serious behavioural changes.

FORAGING BEHAVIOUR

Wild mammals usually spend a significant amount of time foraging. For instance, wild orangutans *Pongo pygmaeus* spend about 46 per cent of their time feeding, wild gorillas about 45 per cent and chimpanzees *Pan troglodytes* about 74 per cent. During foraging wild animals face decisions about where, when, and how long to forage, and which items to choose. However, most captive mammals are deprived of such experiences and are frequently provided with uniform diets presented at the same time and place every day. At one facility foraging time in captive gorillas was reported to have increased from

11 per cent to 27 per cent after the provision of browse; another study on captive gorillas reported about 8 per cent of time spent on feeding and foraging.

It is likely that severely reduced levels of foraging behaviour are prevalent for most captive mammals when compared with their wild counterparts, and variation of food presentation over space and time may allow increased foraging activity in captive specimens. Foraging may be associated with developmental, parental, and sexual behaviour for many species. Provision of more varied diets and the possibility to exhibit natural foraging behaviour may therefore be important for breeding success. Breeding success may also be influenced indirectly by enhancing the general well-being of the animal or by directly stimulating sexual activity. For example, exhibition of more complex feeding behaviours decreases stereotypic behaviours and increases general activity and well-being in different species such as macaques, orangutan several other primate species, and leopards. In the cheetah the stimulation provided by catching and killing live prey appears to enhance breeding success in some instances. Hediger noted that feeding wild animals in captivity should meet nutritional requirements but also needs to fulfill psychological and behavioural requirements.

COMPARATIVE PSYCHOLOGY AND ETHOLOGY

Psychology and ethology of biology

Comparative Psychology and Ethology are both sciences which study animal behavior, typically non-human behavior, though both have often studied humans. Comparative Psychology is a sub-discipline of psychology and ethology of biology.

Both can trace their roots to the late 19th century. Depending on which history one reads the first comparative psychologist was Pierre Flourens, a protégé of Baron Cuvier or George John Romanes, a friend and student of Charles Darwin.

Flourens' book title represented the first use of the term, Comparative Psychology (*Psychologie Comparée,* 1864) and pre-dated Romanes' *Animal Intelligence* (1882). Both proposed a science which would compare animal and human behavior, Romanes postulating the existence of a gradient of mental processes and intelligence from the simplest animals to man - the comparative approach much in use today. Romanes strengthened his proposal by a vast collection of anecdotal accounts of clever behavior in dozens of animal species.

Though perhaps best known today for the fallacies of his anecdotal method and for his easy assignment of human mental faculties to animals—anthropomorphism—Romanes nevertheless succeeded in establishing his idea of a gradient of mental processes across the ani-mal kingdom as a basic premise of early comparative psychology. Ethology too has a mixed parentage. Isidore Geoffroy-Saint-Hillaire first used the term in 1859, though Oskar Heinroth, a late 19th century German biologist was one of the first to apply the methods of comparative morphology to animal behavior; he is thus considered to be one of the founders of ethology.

Both disciplines had many adherents in the early and middle parts of 20th century: Comparative Psychology in the United States under the influence of the learning psychologists (e.g., Ivan Pavlov, Edward Thorndike), the behaviorists (e.g., Zing-Yang Kuo, John Watson, B. F. Skinner), and the epigeneticists (e.g., T. C. Schneirla, Daniel Lehrman, Ethel Tobach, Gilbert Gottlieb); while Ethology became firmly established after WWII in Europe under the influence of biologists such as William Thorpe, Nikko Tinbergen, and Konrad Lorenz.

The latter two, in fact, were awarded the Nobel Prize in medicine (there is no separate prize for behavioral research) in 1972 for their animal behavior studies (they shared this prize with Karl von Frisch, an early 20th century biologist).

Theoretical Background

Given the biological roots of both comparative psychology and ethology, evolution was seen to play an important role in behavioral origins by both disciplines, though in different ways. Comparative psychology, strongly influenced by early 20th century Functionalists (e.g., William James, John Dewey), believed behavior allowed organisms to adapt to their environments (i.e., Darwinism); behavior itself was not an evolved phenomenon, though the organism was.

Thus, as organisms changed through evolution, new or different behavioral potentials arose. Ethologists, on the other hand, understood behavior itself to be an evolved process, the route being genes—instincts, or inherited behaviors. In later years this one-way route, from genes to behavior, became to be known as the Central Dogma of Molecular Biology.

Additionally, while comparative psychology tended to engage primarily in laboratory research, ethology emphasized the significance and importance of studying behavior outside the laboratory, in the natural settings.

These two fundamentally different approaches to the study of behavior lead to a serious intellectual and theoretical ¯war around the 1950s. Ethology advocated the position that behavior was a biological phenomenon, determined, and not merely influenced by the organism's genotype; much animal behavior was thus believed to be in-stinctive. Indeed, Lorenz, whose mentor was Oskar Heniroth, and Tinbergen spelled out the full meaning of what instinctive behavior was.

The clearest statement of this is found in Tinbergen's book, *The Study of Instinct* (1951), Comparative psychologists, on the other hand, tended to take an epigenetic approach, stressing the importance of development, experience, and other psychological proc-esses. The differences were summarized in an important paper by Daniel Lehrman (1953), which today still represents one of the best critiques of instinct theory. While healthy, the ensuing debates settled little. It was an important 1966 book by Robert Hinde (*Animal behaviour: A synthesis of ethology and comparative psychology*) that seemed to resolve the differences between these two opposing views. Indeed, a later 1981 book by the ethologist S. A. Barnett (Modern ethology: The science of animal behavior) was able to discuss the discipline without resorting to instinct explanations.

Important Scientific Research and Open Questions

The two disciplines historically sparred over the nature-nurture issue: Was behavior a biological or a psychological phenomenon? Endless debates over this issue have yet to see it formally resolved. Contemporary reports of the discovery of a gene for a behavior are routinely retracted following failures to replicate such findings - but the search continues.

This is as true in psychology as it is in biology, though many in both camps understand behavior to be a biopsychosocial phenomenon. The significance of both psychological and biological development, long ignored, is now seen to be crucial to a full understanding of behavioral origins. And, though studied now for well over 100 years, there are still new developments to be found in the area of learning.

Current Status

While Comparative Psychology grew in America, ethol-ogy remained somewhat stagnant in Europe. Many still identi-fied with the discipline, though it was clear that they had abandoned the hard nosed biological determinism of the classical ethologists. Beginning in 1944 with the initiation of the Ameri-can Psychological Association's divisional structure, comparative psychology had a home in Division 6, Physiological Psychology and

Comparative Psychology. In the 1990s, in an effort to attract new members, the division entered into discussion of a name change - the important point for the present discussion was the retention of ˉcomparative psychology in the new name adopted at the 1995 APA meeting, Behavioral Neuroscience and Comparative Psychology.

While membership in Division 6 was falling, Comparative Psychology as a field of study remained healthy as illustrated by the appearance of several comparative psychology societies in the clos-ing years of the 20th century: The Southwestern Comparative Psychology Association (founded in 1983 by Michael Domjan, Del Thiessen, Steve Davis and Gary Greenberg); the Comparative Cognition Society (founded in 1994 by Ron Weisman, Mark Bouton, Marcia Spetch and Ed Wasserman; and the International Society for Comparative Psychology (founded in 1983 by Ethel Tobach and Gary Greenberg).

An even earlier group, the International Society for Developmental Psychobiology, was founded in 1967 by George Collier, Norman Spear, Bryon Campbell, John Paul Scott and others. The an-nual and biennial meetings of these societies attract animal behavior researchers from several disciplines across the globe; their membership is also international. There are, of course, several other such societies in countries around the world.

The picture was not so rosy for ethology which seemed to languish in the same period. This was likely because,ˉThe simple truth is that ethology never did deliver as a science of comparative behavior...(Plotkin, 2004, p. 105). Indeed, in 1989 ethology was declared...dead, or at least senescent.

That is, if you think of ethology in the narrow sense – the study of animal behavior as elaborated by Konrad Lorenz, Nikolas Tinbergen, and Karl von Frisch. It has been quiescent for some time. No exciting ideas were emerging, and data gathering on key issues had lost its direction... [Barlow, 1989, p. 2].

However, the biological study of animal behavior has thrived well into the 21st century. Ethology was reborn in the early 1970s as a new science, that of Sociobiology (Wilson 1975), the goal of which was to biologicize the social sciences. But this blatant attempt at understanding animal and human behavior as a purely biological phenomenon was met with scathing criticism (Hull, 1988; Lustig, Richards & Ruse, 2004) from numerous quarters.

The main point of contention centered around the continuing nature/ nurture issue and the question of whether behavior, especially human behavior, was the result of genetic and biological determinism. To many opponents of sociobiology, psychology was not a biological science at all, but a uniquely psychological science (e.g., Greenberg, 2007). The intellectual sparks flew for years, well into the end of the 20th century which witnessed the appearance of a still new iteration of ethology, Evolutionary Psychology. This approach focuses primarily on human behavior and posited that we owe our universal nature to evolutionary adaptations faced by our Pleistocene ancestors that we have inherited in our genomes. A good source for reviewing the tenets and the research conducted in this field is The Handbook of evolutionary psychology (Buss, 2005). With evolutionary psychology, instincts are once again in vogue. As with ethology and sociobiology, evolutionary psychology is not without its critics (e.g, Honycutt & Lickliter, 2003). It is not the application of evolution to behavior that is at question, but the manner in which it is understood to apply to behavioral origins. Evolutionary Psychology, though seen by many to be seriously flawed, is a rather popular orientation in the contemporary behavioral sciences. After all, what serious scientist in 2010 can object to the significance of evolution to psychology?

There has also been new life breathed into ethology and sociobiology. The sociobiological idea of the genetic basis of human altruism has recently been somewhat retracted by one of its earliest proponents, E. O. Wilson. While this is comforting news to many non-reductionistic comparative psychologists and other animal behaviorists, it doesn't sit well with all students of behavior (Marshall, 2010) attesting to the staying power of the classical ideas of ethology. In a recent analysis Salzen (2010) makes a case for interpreting the ideas of ethology in modern neuroscientific terms.

There is in fact a discipline known as Neuroethology, which describes animal behavior in terms of how the nervous system works. As a comparative psychologist, I take comfort in the staying power of my discipline. Its history has been long, though not nearly as tumultuous as that of ethology.

METHODS IN BEHAVIOURAL ECOLOGY

Behavioural ecology focuses on the adaptive significance of variation in individual behaviour, morphology, and physiology which has forced scientists to learn to recognize individual animals.

Consequently, all the individuals within a subpopulation are often known to researchers fortuitously providing an accurate estimate of local population size. Individually recognizing all members of a population is a far more accurate method of estimating population size than using classical ecological techniques such as mark-recapture methods or stratified sampling. In addition, long-term field studies have unusually accurate figures of changes in population size in turn, this may influence effective population size. McGregor and Peake review methods used to recognize animals individually and additionally show that populations can be monitored through individual call recognition rather than observing animals themselves.

MOLECULAR TECHNIQUES

In studies of known individuals in the wild, data are regularly collected on paternity, and sometimes maternity, using genetic markers. These data generate information on the extent to which particular individuals contribute to the next generation and hence to N. Knowledge of the number of lineages, or breeding groups, for a given Ne will alter estimates of the time frame over which allelic diversity is maintained. Allozymes, DNA fingerprinting, mitochondrial DNA (mtDNA), minisatellite, and microsatellite techniques shed light on population differentiation and genetic variability.

These methods have identified subpopulations of humpback whales *Megaptera novaeangliae)* migration routes and extent of philopatry in marine turtles, and the degree of hybridization between canid species information almost impossible to obtain from observation alone. In addition, analyses of mtDNA sequences determine phylogenetic relationships between species, which help identify evolutionarily significant units (ESUs) for conservation. For instance, genetic analyses of orangutan *Pongo pygmaeus* populations indicate sufficient divergence between Sumatran and Bornean populations to put a halt to breeding individuals from the two regions. Although within-and between-species molecular comparisons are normally outside the purview of behavioural ecology, analyses that attempt to uncover associations between behavioural and morphological variables without the confounding effects of shared ancestry nonetheless require phylogenetic trees derived using molecular techniques.

HUMAN BEHAVIOURAL ECOLOGY

Because anthropogenic factors are the greatest challenge to biodiversity, many conservation biologists would argue that conservation solutions will eventually emerge from understanding human behaviour. Patterns of human exploitation are better understood now than they were 15 years ago. The myth of the "noble savage" living in harmony with the environment has been exploded.

In a detailed exploration of this issue, Alvard uses empirical data to show that Piro hunters in Peru pursue an optimal foraging strategy rather than any conservation agenda. As a second illustration, the tragedy of the commons, which describes individuals overexploiting a common property, is better understood in terms of a problem of unrestricted access rather than a problem of ownership. This focuses attention on the best ways to police resources rather than on who owns them.

These findings are informative in that they show which conservation strategies are likely to succeed or fail. In spite of these advances, the rapid growth of evolutionary psychology, a subdiscipline in which contemporary human behaviour is interpreted in the light of past selection pressures, has so far had little to say about patterns of current exploitation. Similarly, economics does not consider the underlying biological motivation in decision making.

Nevertheless, it seems likely that we will be able to understand human resource use better by paying closer attention to evolved behavioural dispositions. Specifically, they draw attention to differences in attitudes towards conservation shown by different segments of society as predicted by evolutionary theory. Certainly, the two pivotal problems in conservation biology, human population growth and resource accrual, are central to human behavioural ecology's focus on the social and ecological factors that affect reproductive success in different cultures. Demographic records and information on dispersal are obviously of great importance, while population growth rate, a population's response to exploitation, and Ne are influenced by many factors in behavioural ecology.

There are other bridges between these disciplines that have not been included, however, and many others that doubtless will appear as our knowledge of behavioural adaptations improves.

ETHOLOGY AND STRANGE ANIMAL BEHAVIOR

Ethology is the study of animal behavior, based in zoology. Ethologists study the evolutionary basis and development of animals' innate behaviors, like a spider knowing how to make a web without learning from a parent.

They also study forms of communication (physical, chemical, visual) and social interactions between animals. Human ethology research looks at the evolutionary origins of human behavior and also compares behaviors across cultures. Other studies of animal behavior are based in psychology, focusing on things like learned behavior and teaching behaviors to animals and applying the results to humans. One common question about animals that can be considered through the lens of ethology is whether animals have special sensing abilities. For thousands of years, stories have spread about animals sensing earthquakes.

Just before the 2004 tsunami that ravaged parts of southern Asia, many animals exhibited strange behavior or ran (or flew) to higher ground. By some accounts, rescue workers found a surprisingly low number of dead animals, though there were areas where many dead animals, particularly cattle, were discovered. Is this another case of a special "sixth sense" or supernatural ability? Some scientists propose that sophisticated hearing and the ability to detect subtle vibrations allow animals to sense earthquakes.

Some also suggest that animals detect changes in the air or in electromagnetic fields. In any case, it's likely not a mysterious sense but rather one or two senses — such as hearing and smell — that are so highly refined that animals can hear an earthquake or smell gases released by that earthquake. Alan Rabinowitz, of the Wildlife Conservation Society, claims that humans once had this ability but lost it through evolution [Source: National Geographic]. (CBS' "60 Minutes" produced a remarkable story explaining how the seafaring Moken people used their close connection to the ocean to detect the tsunami before it happened.)

It's very difficult for scientists to pin down exactly what causes animals to flee or panic before an earthquake. Numerous such stories exist, but a reliable testing method does not. After all, animals respond to many stimuli, some of which are difficult to trace. Critics also say that people remember their pets' acting differently only because an earthquake happened, and that pets often exhibit strange behaviors with no subsequent earthquake or disaster. But the theory of animals' sensing earthquakes has gained enough ground that scientists around the world have attempted to test it, with mixed results.

Not only do some animals have highly refined senses, but many experience sophisticated emotions as well. Frans B.M. de Waal, a noted primatologist at Emory University, says that a variety of animals — not just catsor dogs, but even rats — feel empathy and other emotions.

In one study, scientists injected mice with a chemical that gave them slight stomach aches. The natural behavioral response is stretching, and injected mice stretched more when placed next to injected mice than they did when placed next to non-injected mice.

Male mice also showed less of a response around males that they didn't know. In other words, not only do mice show a response to the pain of others, but it matters who the other mouse is. University of Chicago neurobiologist Peggy Mason called this "a significant step toward human-like social feeling".

BENEFITS OF SOCIAL BEHAVIOUR

Social behaviour seems to provide many benefits to those who practice it. Studies have shown that many animals are more successful in finding food if they search as a group. This is especially true if food resources are clumped together only in certain places.

If more individuals are cooperating in the search, there is a greater chance one of them will find the clump of food. In some cases, foraging in a group makes it easier to capture a prey.

Dolphins are known to surround a school of fish and to take turns darting into the center to eat the fish that are trapped in the middle. Many carnivores will band together when they try to capture large prey. For examples, wolves will hunt together when hunting moose, and lions will hunt together when hunting large prey such as wildebeests. When these animals are hunting much smaller prey, they will often hunt singly.

*Fig. **Two Gannets Greeting Each Other on Bonaventure Island, Quebec.***

Many animals live in social groups partly for protection. Although one baboon might not be able to fight off a leopard, a troop of baboons often is able to do so. In addition, with more individuals cooperating together, some can serve as sentries looking for danger while the other group members are eating or sleeping. Prairie dogs and large flocks of crows normally have some individuals acting as sentries, which makes it nearly impossible to sneak up on a prairie dog town or a flock of crows.

Many prey species, such as schools of fish and flocks of shorebirds, travel in groups in which their movements are highly coordinated. The entire group moves quickly, darting one way and then another as an entire group, as if they were all somehow physically connected with one another. It is believed this behaviour creates confusion for the predator. Predators generally need to pick out a single individual in a group that they will focus on and try to capture. A rapidly moving and turning school of fish, flock of birds, or herd of antelope is believed to make it very difficult for the predator to remain focused on a single individual. However, if one individual is unable to keep up with the group, the predator will then be able to focus on it and usually will succeed in catching it.

Some animals form social groups to make travel easier. Canada geese and other bird species typically fly in a *V* formation. Just like bicyclists who ride behind one another in order to reduce wind resistance, the geese fly in formation to reduce the wind they must encounter. In this situation, the lead bird has the most tiring job, which is why several birds usually take turns leading the *V*. Some animals congregate in close proximity to one another in cold weather in an effort to stay warm. Small birds are sometimes known to huddle so closely they form a single large ball of birds.

Breeding Behaviour of Groups

Sometimes social behaviour is exhibited by groups of males or females during the breeding season. In some cases, males may band together and try to chase the dominant male away so they have a better chance of mating success. In other instances, males are known to cooperate in making their courtship displays.

Turkeys often perform their courtship display in pairs, even though only one of the turkeys ends up doing most of the mating. Why would the unsuccessful male agree to help? The two male turkeys are usually brothers. Since brothers share about 50 percent of the same genes, even if only one brother mates, many of the genes of the unsuccessful brother are passed on too.

In some species, the females form social groups during the breeding season. In certain circumstances, females will look after one another's offspring while the other mother goes out to find food. In other species, such as lemurs, females may form social groups as a kind of defence. Males of some lemur species will try to kill the offspring of females that mated with another male. By banding together, the females are sometimes able to ward off the attacking male. Many animals form social groups only during certain times of the year. Many bird species flock together in foraging groups in the winter. However, these same birds that sought one another out in the winter set up breeding territories in the spring and will go to great lengths to keep the same birds out of their territory. Thus, for many species, social behaviour is a flexible form of animal behaviour, one that can be adopted or abandoned depending on the conditions of the environment and the time of year.

Developed of Insect Societies

Some of the most well-developed social behaviour is exhibited by insects such as ants, termites, bees, and wasps. Many of these species live in colonies with thousands or even millions of individuals. One benefit of social behaviour for these insects is that different individuals specialise in certain activities. For example, some are the workers who build the colony and go out looking for food that they bring back.

Other individuals are the soldiers of the colony. Their job is to continually patrol the colony perimeter and to protect the colony from possible attacks from other colonies. In many ant and bee colonies, all worker and soldier ants are females.

Males are usually present in the colony, but do not contribute much. Finally, there is the queen ant or bee. The queen's only job for her entire life is to lay eggs that the workers will care for.

Fig. ***An Olive Baboon Grooming Another in Nairobi National Park, Kenya. Individuals Who Engage in Social Behaviour are More Likely to Survive and Reproduce.***

There are substantial benefits to forming social groups and there are also some definite costs to living closely with others of the same species. First, one competes most with others that are most like oneself, and thus a member of a social group always has to share or compete with others for resources. Second, because of the numbers and close proximity of individuals in many social groups, disease may spread through social groups relatively rapidly.

Distribution of Social Behaviour

The parame-cium, so commonly studied in elementary biology courses, is a highly developed protozoan which exhibits a variety of behaviour patterns. One of these is a simple form of sexual behaviour in which the paramecia fasten themselves together and afterwards exchange nuclei and divide. Another pattern occurs under unfavourable environmental conditions. When a drop of water containing several paramecia is allowed to dry slowly, the animals will huddle together and thus protect each other against drying out. Such behaviour has the function of shelter or comfort seeking. In addition, this one-celled animal shows other patterns of behaviour which are only incidentally social. One of these is an investigatory or exploratory pattern of behaviour in which the animal drives itself ahead with its cilia in a spiral path and thus comes in contact with various objects. Depending on the nature of the objects, it may remain in contact or back off and start out in another direction. Finally, its pattern of ingestive behaviour is extremely simple, consisting of driving water laden with food particles towards its gullet by means of cilia, and it involves only starting or stopping the beating cilia.

More complex patterns of behaviour having basically the same functions are present in the lower invertebrates, but only in the arthropods and vertebrates can we find new functions. Male crayfish, for example, will fight with each other during the breeding season, an example of agonistic behaviour. Certain insects show highly developed patterns of behaviour for the care of the young, thus exhibiting care-giving, or epirneletic, behaviour. The young larvae, by their movements, may signal for care and attention, exhibiting care-soliciting, or et-epimeletic, behaviour. Likewise, ingestive behaviour assumes a highly social function, especially among ants, which feed each other in a relationship that Wheeler called trophallaxis.

All of these behavioural functions are highly developed in vertebrates. In addition, superior eyesight permits vertebrates to develop patterns of social behaviour which involve mutual imitation; this allelomimetic behaviour is commonly seen in schools of fish, flocks of birds, and herds of mammals, where each animal follows the movements of those next to him, resulting in highly coordinated group movements.

Systems of Behaviour

As behaviour becomes more complex in the animal kingdom, a given species develops several alternate patterns of behaviour for each general function. A paramecium, for example, has only one simple pattern of ingestive behaviour and no patterns of agonistic behaviour. In contrast, a mouse attacked by another mouse may fight back, run away, assume a defensive posture, or become limp and passive.

Each of these patterns of behaviour has the function of adapting to the same stimulus, and the attacked mouse usually goes through all or part of this behaviour repertory until one response is finally selected on the basis of its effect. Such a group of behaviour patterns having a common general function, and usually a common physiological basis, is a behavioural system.

Behavioural systems are quite similar to physiological systems, and many of them are directly related, as for example, sexual behaviour and the reproductive system.

Like physiological systems, behavioural systems do not function entirely independently of each other, and in various species they may be organised on a somewhat different basis. In the dog and other members of the family Canidae, sexual and eliminative behaviour are closely related to each other, whereas in birds the two are unrelated.

The nature of behaviour patterns and their organisation into a system is largely a result of the special heredity of the species concerned. A primary task in studying the social behaviour of any species is to describe its special behaviour patterns and their organisation into behavioural systems. This behavioural inventory, or ethogram, as Tinbergen has named it, provides the essential raw material for understanding social organisation or conducting experimental work.

Expansion of Behavioural Systems

In most of the higher animals, behavioural systems are incompletely developed at birth or hatching. In young mammals the ingestive system is present in an immature form, the newborn animal taking its food only by nursing. The development of behaviour patterns for ingesting solid food comes later. Some systems of behaviour may be completely absent, and all of them tend to be loosely organised together as et-epimeletic behaviour. As the animal grows older, both behaviour patterns and behavioural systems become differentiated from each other, partly as the result of learning and partly because of the processes of growth and maturation.

Social Organisation and Relationships

A social relationship may be defined as regular and predictable behaviour between two individuals, usually of the same species. In some animals these relationships may be organised largely as the result of heredity, but in higher animals relationships develop as a combined result of hereditary factors and learning. When two chickens meet for the first time, each presents a problem of adaptation to the other, and one solution is to attack. The result is a fight, and usually one chicken wins and the other loses. When they meet again, there is a tendency to repeat the behaviour, so that in the long run one individual forms a strong habit of winning and the other of losing.

Since one of the principles of learning is the "law of effect" (which might also be called the law of least effort), the winning chicken reduces its attacks as time goes on, and the whole relationship is reduced to one of threat and avoidance behaviour rather than actual fighting. This is an example of a typical dominance-subordination relationship.

Hereditary factors also help to determine the development of the relationship. If one bird is a male and the other a female, the male almost invariably wins. Among females, large hens are likely to win over small ones, and aggressive breeds over peaceful ones.

Thus, the development of a social relationship may be considered as a process of differentiation of behaviour, affected by both heredity and learning. As will be seen below, the development of social relationships in young animals is a slower and more complex process.

Social relationships can be classified in two different ways. In Carpenter's system, classification is based on the different kinds of individuals present in the system. In howling monkeys and most other mammals there are three major kinds of individuals (males, females, and young) combined in six kinds of social relationships. The same scheme applies to all social vertebrates, but it becomes more complex in some of the social invertebrates. In bees there are three kinds of adults—males (drones), females (queens), and sterile females (workers)-as well as the young. In addition, the workers exhibit different behaviour at different stages of development. In termites there may be as many as three different kinds of individuals capable of reproduction and as many more different kinds of workers.

Social relationships may also be classified according to the behavioural system or systems involved. On this basis a large number of social relationships are theoretically possible, but actually observed, important social relationships are relatively few in number.

There are four which involve one behaviour system only. Dominancesubordination relation-ships are developed in any animal system where agonistic behaviour is found. Sexual relationships depend of course on sex behaviour. Leader-follower relationships are based on allelomimetic behaviour and result from an unequal degree of mutual imitation. Finally, mutual-care relationships, often seen in the mutual grooming of primates, are formed when both animals exhibit epimeletic behaviour.

There are two important relationships composed of combined systems: trophallaxis and care-dependency. Trophallaxis involves mutual feeding and mutual care and is very important in the social insects. The workers feed the young and in turn lick off secretions from them. In mammals, birds, and other vertebrates which care for their young, epimeletic and et-epimeletic behaviour are combined to produce the relationship of care-dependency.

Both of these schemes of classification may be combined for systematic study. For example, the behaviour exhibited in a male-male relationship may be examined in detail and classified as dominance-subordination, leader-follower, or whatever it may be in the particular animal society.

The number of possible social relationships which an animal develops depends on the size of the group. Carpenter has presented a simple formula for the total number: $n(n-1)/2$, which is of course the number of possible combinations between any two individuals in a group of n individuals. When examined in more detail, this formula has certain interesting properties which have implications for the social organisation of groups.

If we consider groups of different sizes, starting with one and going up to ten, we see that an individual added to a group adds a number of relationships equal to the number of individuals already present, *i.e.*, an individual joining a group of three adds only three new relationships, but adds ten to a group of ten. There is a very rapid increase in the number of relationships as the group gets larger, resulting in extremely complex organisation. Experiments indicate that hens in a flock of one hundred can recognise every other individual, making a theoretical total of 4,950 relationships in the flock. Social activity tends to increase in proportion to the number of social relationships.

Primary Socialisation

In highly developed societies the individuals must become related to particular members of the species. An ant colony would become totally disorganised if its members reacted towards members of other colonies in the same way as to their own. Ants would wander from one nest to another, overcrowding one and leaving young unprotected in others. Therefore, at some period in their development, all individuals must learn to differentiate their own family or colony from others. As Fielde showed, there is a critical period in early development in which the immature ant can be transferred from one nest to another or even to that of another species and subsequently act as if it belonged with the new individuals. This transference of social relationships occurs regularly in the slave-raiding ants, a species of red ant commonly found in North America. The raiders remove the pupae from nests of black ants and take them to their own, where the black individuals subsequently rear and care for the offspring of the raiders.

A similar phenomenon takes place in all highly social animals which have been studied so far. Birds hatched in incubators or removed from their nests in early stages of development and reared by hand become attached to the human species and often fail to react to their own. In precocial birds, such as chickens, ducks, geese, and turkeys, the formation of the first social relationship takes place within the first 24 hours after birth. In the more slowly developing altricial birds, such as song sparrows, jackdaws, or doves, the process takes place at a later period in development, and the degree of attachment to the human handler varies inversely with the age at which the young are taken from the nest.

Among mammals, also, the process varies with speed of development. In sheep and other herd animals the young lamb becomes strongly attached to its mother within the first few days, whereas in the more slowly developing dog the period of socialisation begins at about three weeks of age, reaches a peak at about six or seven weeks, and slowly declines thereafter.

The process of socialisation is of course a reciprocal one, the parents becoming attached to their offspring as well. In sheep or goats the mother becomes attached to her offspring within the first two hours or so and thereafter will repulse any strange infant. Primary socialisation has been little studied in primates other than man, but it takes place within the first six months in rhesus monkeys, as also appears to be the case in human infants.

The time at which primary socialisation takes place is obviously a critical period in development, determining which individuals shall be the close relatives of the young animal. Later he may exhibit appropriate patterns of sexual behaviour and other types of social interaction towards similar individuals.

What is the nature of the process involved in primary socialisation? In the young animal there are positive behavioural mechanisms such as the"following" reaction of a young chick or lamb, which brings it into contact with members of its own species. There are also negative mechanisms, particularly the behaviour patterns of escape, which develop somewhat later and prevent socialisation with other individuals by inhibiting contact. As to the nature of the process itself, in precocial birds it obviously takes place in the absence of food rewards, and, under experimental conditions, without any obvious external rewards or reinforcement of any kind.

While food rewards do affect the behaviour of a young mammal, experiments have shown that the process of socialisation takes place without them. All evidence indicates that the process is an internal one and is closely connected with emotional responses. Experimental studies indicate that any kind of strong emotion, whether it be that of loneliness, fear, pain, or hunger, will speed up the process of forming a social bond. While the capacity to form such a bond is strongly developed in the infant animal, the same process can take place at any time in later life at a somewhat slower rate, except when it is completely suppressed by the development of interfering behaviour patterns.

ANIMAL AND HUMAN SOCIETIES

The kind of social organisation developed by a species depends upon the nature of its behaviour patterns and the ways in which they are organised into behavioural systems. Thus we may state a general law: Social behaviour is a determinant of social organisation.

Aggregations

The simplest kind of animal society is the temporary aggregation. Such groups occur widely in the animal kingdom and were extensively studied by W. C. Allee, who was able to show in a large number of cases that living in groups promotes the survival of individuals, whether they are protozoans or vertebrates. This kind of temporary aggregation is based on shelter–or comfort-seeking behaviour.

A second kind of temporary aggregation is based on sexual behaviour. This may involve only two individuals, as it does in paramecia, or vast numbers, as in the mating swarms of the palolo worm of the Pacific. The great majority of invertebrate societies belong to one or the other of these temporary aggregations.

In contrast, the higher invertebrates, and many vertebrates, form long-lasting groups with highly complex internal organisation. Insect groups first received scientific attention, but it is now realised that the social behaviour of insects is somewhat specialised, being limited by certain anatomical peculiarities such as the external skeleton and the relatively inefficient compound eyes. Because of these limitations, insects are small, relatively shortlived, and unable to respond to complex visual stimuli at a distance.

Insect Societies

The most highly developed insect societies belong to two orders, the Hymenop-tera, including ants, bees, and wasps, and the Isoptera, or termites. In both orders the societies consist of permanent groups associated with nests that are constructed by the combined efforts of the members, and the most prominent social behaviour is care giving. In an ant colony two workers meet, feel each other with their antennae, and if one has recently fed, it will regurgitate a drop of honey dew to be eaten by the other. This, together with care of young, constitutes most of their social behaviour and is what Wheeler termed trophal-laxis.

In any given species of insect, social organisation tends to be rigid and stereotyped, repeating itself generation after generation. There is relatively little modification by experience, and the combined efforts of the individuals produce a stable social environment. Alfred Emerson points out that insect societies have many of the characteristics of a single organism, and he has called them "supra-organisms". The peak of development of the insect society is reached by the common honeybee, which maintains a permanent social group. In contrast to ants, in which each new colony is formed by a mated pair, large numbers of old bees accompany a queen moving to a new hive. Mating takes place at this time, and the fertile female, or queen, continues to reproduce for a long period, sometimes two or three years. The vast majority of the colony members are short-lived sterile females or workers which go through a regular series of activities as they pass through a few weeks of active life, first working around the hive and finally going out to forage for food. As von Frisch discovered, foraging bees are able to communicate to other workers the direction and distance at which food is found by means of body movements, or "dances," when they return to the hive (1950). Present evidence indicates that the nature of the signals and the ability to understand them are determined by heredity rather than learning.

Vertebrate Societies

The societies of vertebrate animals are also related to basic anatomy and physical capacities. Vertebrates have an internal skeleton and hence are capable of continuous growth. They can become immensely large, and their life spans may be enormously extended incomparison with insects. The capacity for learning is consequently much more valuable and plays a more important role than it does among invertebrate societies, so that there are many examples of cultural as well as biological inheritance. Furthermore, the vertebrate eye is vastly more efficient than the compound eye of arthropods, making mutual imitation or allelomimetic behaviour possible, particularly for animals active in the daytime.

There is an enormous variety among vertebrate societies, ranging from temporary aggregations, like those of the lower invertebrates, to societies far exceeding those of the insects in complexity. The most highly developed societies are found among animals which occupy dominant ecological positions.

In the water, fish are still the dominant form of life, and their societies have taken two forms. One of these, the school, which is based on allelomimetic behaviour, is very widely found. In herring or mackerel the fish move in groups throughout their lives, even spawning in a mass and paying no attention to the eggs or young. In contrast, there are many kinds of fish, like the sunfish and stickleback, which emphasise nest building and the care of the young. In these cases the male guards the nest and sometimes the young for a short period after hatching. The period of dependency remains short, however, as in most cases the young must feed themselves. The males guard definite boundaries around the nest and thus exhibit territoriality.

Amphibians, which usually occupy a relatively inferior ecological niche and are under constant pressure for survival, show almost no social organisation except temporary mating aggregations and a few examples of care of the young. The same can be said for many reptiles, although some lizards develop dominance and territoriality and alligators build nests and briefly protect their young.

Birds, which are the dominant form of life in the air, show a great variety of highly developed societies. One of the characteristics of bird societies is the phenomenon of seasonal change. For example, during their brief mating season, song sparrows develop one sort of social organisation involving sexual behaviour, territoriality, and care of the young and then spend the rest of the year in flocks which are much like schools of fish. Since the young are produced as eggs which have to be incubated, a high degree of coordination of care-giving behaviour and physiological processes is necessary, and in many birds this is exhibited by both sexes, which combine to build the nest and care for the young. Many species feed their young, and pigeons and other members of the dove family nourish their offspring with crop milk. A "nuclear family" of male, female, and young is characteristic of these and many other bird societies, but in chickens and other members of the same family, mating is polygamous, and the brood of young is raised by the female alone.

In mammals there is an enormous variety of societies differing in structure and degree of complexity. Among rodents, woodchucks are relatively solitary animals, reacting to each other with agonistic behaviour except during the brief mating and reproductive seasons. At the opposite extreme are the prairie dogs, which run to groups numbering thousands of individuals. Each prairie dog colony is subdivided into territories, usually occupied by a sexually mature male, a group of females, and their young. Each year the adults in the territory move out to new locations on the edge of the colony, leaving the burrows behind for the next generation. Thus the oldest and most experienced members of the colony are found on its edges, where adaptation is more difficult.

Other examples of mammalian societies are the great herds of hoofed mammals. In a typical society, such as that of mountain sheep, the two sexes mingle during the mating season. Males compete in individual combat, and the more successful ones do most of the mating. During the rest of the year there are separate male and female herds, the latter being the group in which the young are raised. Among these herds a definite leadership organisation develops, the older animals of either sex tending to lead in their respective groups.

Some carnivores are quite solitary in their habits, but others, like wolves and dogs among the Canidae and lions in the cat family, live in permanent groups. These are, however, much smaller than those of the herd animals on which they prey.

Thus, we see that there is an enormous number of different kinds of animal societies. There is great variation among human societies, but the variety possible in other animals far exceeds it. It is therefore of considerable interest to see what kinds of social organisation are present in man's closest biological relatives, the other members of the Primate order.

Primate Societies

Like other important orders of mammals, the primate group includes a wide variety of species, ranging from the small "bushbabies," or galagos, of Africa to the large manlike apes and man himself. Primates can be at least as different from one another as cats are from bears in the order Carnivora. Their chief limitation of variation is an ecological one. Except for man, they live entirely in tropical and subtropical regions, and while many primates like to play with water, none of them is truly aquatic.

That primates other than man are highly social was first established by Carpenter's classical study of the howling monkey. These are tree-dwelling primates of Central America. A band of males, females, and young wanders from tree to tree, eating fruits and leaves and rarely descending to the ground. Their most obvious kind of social behaviour is allelomimetic, since they are constantly following each other. The more active males generally take the lead, but there is no single established leader in their wanderings. Agonistic behaviour is greatly reduced and expressed almost entirely as the vocalisation that gives the species its name. Females in estrus pass from male to male, and there is no indication of sexual jealousy or rivalry over females. The most important social relationship within the group is that between mother and offspring, the mother constantly carrying and caring for her baby for the first two years of its life. Care-giving behaviour is also shown by the males if a young animal falls to the ground or is threatened in some way. Different bands have roughly marked territories, and groups keep each other apart by means of threats.

Like man, baboons are a plains-living primate, and for this reason their social life has particular interest. Biologically they have evolved towards a doglike structure, often running on all fours and developing enormous teeth. However, they are largely vegetarian rather than hunting animals, often eating grass which they pick and eat a blade at a time. They readily adapt to living off farm crops, particularly melons and vegetables, and can be enormously destructive.

As with howlers, a large part of their social behaviour is allelomimetic, and they move in a group of males, females, and young which may include thirty to one hundred individuals. Agonistic behaviour is more prominent than in howlers, and the males develop a definite dominance order. As the group moves, the most dominant male stays in the center, while the more subordinate animals keep their distance from him and each other. Living on the plains, the group is constantly threatened by large predators such as lions and cheetahs.

Females and their young tend to group themselves towards the center of the group, near the dominant male, this being the safest spot. When a member of the group is attacked or threatened, all the males combine in its defence. Consequently, baboons are seldom actually harmed by predators. When in estrus, a female may mate with the most dominant male first but then move on to others as he becomes satiated. As with howlers, there is no evidence of sexual jealousy or permanent bonds between particular males and females. The most important social relationship is that between mother and offspring; in fact, all adults are extremely responsive to young animals and will approach and care for them if the mother permits. As the infant becomes more independent, it joins a play group of other young animals within the group. In general, the baboon society gives the impression of a closely knit and highly cooperative group.

Macaques are related to baboons, and their behaviour is somewhat similar. The Indian macaque, or rhesus monkey, commonly lives near human habitations and is often fed by people. Carpenter studied an artificial colony of rhesus monkeys on Santiago Island off Puerto Rico and demonstrated that a dominance order between the males was an important part of their social life. As with baboons, females in estrus pass from male to male. Overt aggressive behaviour is perhaps more common than in baboons, possibly because of the disorganising effect of human contact. Southwick found that where rhesus troops went into cities and competed for food, there were many cases of serious injury.

The behaviour of the Japanese macaque is quite similar. Imanishi reports that the animals spend their lives within a troop composed of males, females, and young. The most dominant male usually occupies the central position, surrounded by females, while the younger and less dominant males keep to the outside of the troop and also precede it when it is moving. As with baboons, there is no subdivision of the troop into families or any permanent consortship between individual males and females.

The Indian langur, or Hanuman monkey, is a tree-living species and exhibits much less aggressive behaviour than the macaques. However, Jay has described a definite dominance order among males. AH females are subordinate to all males, except when an infant is threatened, and their own dominance order is not a clearly marked one. Sexual behaviour is relatively unimportant, as the females have only a five-day estrus period per month and this is absent during pregnancy and lactation. Females have some tendency to consortwith the most dominant male, but there is no absolute correlation, and in one group studied by Jay the most favoured male was a low-ranking one. Male-female consortships never last more than a few hours. In contrast to baboons, males pay little attention to the young, but infants, and particularly the newborn ones, are the object of much attention from all females in the troop. Males tend to take the lead when the group moves.

The manlike apes have been very difficult to study in the wild. In his original study Nissen was able to see very little besides the retreating chimpanzees disappearing into the tops of the tropical forest. More recently, Kortlandt located a favourable area on a Congo plantation where chimpanzees were protected and were in the habit of coming into the open. Reynolds and Reynolds were able to keep track of a group of chimpanzees in the Budongo forest of Uganda with the aid of a large group of spotters scattered through the dense vegetation. Unlike baboons, chimpanzees do not live in compact groups, but forty to fifty individuals may occupy an area of six to eight square miles. Except for females and their offspring, there are no permanent associations between individuals, and temporary groups of males and females, females and young, males, and groups containing all kinds of individuals are found in approximately equal numbers.

Agonistic behaviour is uncommon, and there are few indications of dominance and subordination. During estrus the same female may mate with several males in succession, and no fighting is observed between them. Chimpanzees appear to keep in touch with each other by vocalisation, and on many occasions groups will join in vocalisation over a large area. There is little evidence of leadership under most conditions and no reports of defence of territorial boundaries. Thus, chimpanzees form a very loosely knit but mutually tolerant society, with loose temporary associations between individuals, except for the mother-offspring relationship.

Goodall has studied chimpanzees in the comparatively open forests of the Gombe Stream Reserve in Tanganyika and has been able to follow the behaviour of identifiable individuals over a period of several years. Her more detailed accounts of behaviour confirm the general picture of chimpanzee social organisation. In addition, she has observed occasional predatory behaviour and meat eating. These chimpanzees also exhibit use of tools, in particular, poking twigs or rolled-up leaves down the holes of termites in order to collect them for food.

The behaviour of gorillas under natural conditions is also surprising. With their immense size these animals have little to fear from predators and within their own social groups are placid and peaceable. They are strict vegetarians, and the two sexes are so much alike that they are almost impossible to distinguish at any distance. Their sex organs are very small, and sex behaviour plays a very minor part in their social life. Unlike chimpanzees, gorillas live in small compact groups containing both sexes and all ages of individuals. According to Schaller, there is a strong system of leadership, all members of the group responding to one of the older males. Dominance is expressed almost entirely in the form of allowing precedence, and it is strongly related to age.

The only one of the manlike apes that shows anything like the human nuclear family of male, female, and their offspring is the gibbon. These long-armed tree-dwelling animals of southeast Asia are highly aggressive in both sexes, and the largest social group is a male, a female, and their immature offspring. Other groups are kept away by threats and fighting. Rather than supposing this to be the beginning of evolution towards a nuclear family, we can advance the hypothesis that the gibbon group is formed by a reduction of the troop to the smallest unit which is still biologically functional.

Primate societies do not present an evolutionary history of human social organisation. Rather, each species has evolved its own social organisation, producing an astonishing variety against which it is possible to compare and contrast human social organisation. The one thing which is constant in all primate societies so far studied is the emphasis on the care of the young, particularly by the mother but also by unrelated males and females in the highly social forms. Agonistic behaviour varies a great deal, as does sexual behaviour. Most primates show regular periods of estrus, although there is some indication of a tendency towards a more extended receptivity in female chimpanzees. Was the social organisation of our primitive human ancestors more like troops of rhesus and baboons, or was there a tendency to form temporary separate groups based on sex and age, as in the chimpanzee? The almost universal tendency towards division of labour in human societies suggests the latter. The animal data also suggest that the human nuclear family is perhaps not the beginning but the end point of human social organisation, resulting from situations which demand great fluidity; that is, the nuclear family represents the ultimate reduction of family organisation to the simplestunit which is still capable of all the functions of biological reproduction.

THE EVOLUTION OF SOCIAL ORGANISATION AND BEHAVIOUR

Social organisation is based on behaviour, and behaviour leaves no fossils. Therefore, any reconstruction of the evolution of animal societies must always be hypothetical and based on what we know about living animals. Assuming that the more lowly organised forms of living animals are similar to those which existed millions of years ago, we can suppose that there are two general bases for the initiation of social life. One of these is shelter-seeking behaviour, in which animals stay together because the bodies of their fellows form a favourable environment. If this sort of behaviour is extended, it should result in the animals' following each other around, *i.e.*, allelomimetic behaviour. Animals that do the latter must have both motor equipment for rapid coordination and the necessary sense organs to keep in touch with each other.

This kind of social group approaches its highest development in schools of fish, flocks of birds, and herds of mammals.

The other kind of social behaviour which we may assume to be primitive is sexual behaviour, which results in more efficient means of reproduction than scattering the sex cells broadcast. Retaining the eggs after they are fertilized and giving them care after they are laid still further increases the efficiency of reproduction, and the height of this kind of behaviour is seen in two widely different parts of the animal kingdom—societies of insects and those of many birds and mammals.

These two basic kinds of social organisation may evolve independently of each other, or both may occur together in the same species. In many cases social organisation is quite independent of physical form, and closely related species may be either highly social or relatively unsocial. For example, the deer family range from the highly social elk through the more solitary Virginia deer to the moose, which is almost completely solitary except during the brief mating season and the association of a calf with its mother. Among rodents there is the prairie dog, a ground squirrel that lives in colonies of thousands of individuals, and at the opposite extreme, woodchucks, which lead highly solitary lives except for the brief necessities of mating and rearing the young. Even in the social insects there are close relatives of honeybees which build solitary nests. Thus there is no consistent relationship between biological form and social behaviour, and the evolution of societies has become largely, though not entirely, independent of the anatomy of the individual. Consequently, in searching for the origin of human sociality, it is not possible to find anywhere the recapitulation of human social prehistory. The study of animal societies can at most suggest ideas and provide a basis of comparison.

The Problem of Basic Human Nature

In addition to biological evolution, human societies have added a new dimension: cultural evolution. The beginnings of this can be seen among the higher animals where the offspring learn fears from the previous generation. However, human language has enormously increased the potentiality of passing information along from one generation to another, and written language permits its accumulation over almost infinite times and in enormous amounts.

Compared to biological evolution, the study of cultural evolution is still in its infancy. Very little progress was made in the theory of biological evolution until the mechanism of heredity was discovered. Reasoning by analogy, we may suggest that the development of an adequate theory of cultural evolution will in part depend on our knowledge of the mechanism of cultural heredity, namely learning, and its application to cultural phenomena. Unlike biological evolution, which moves in terms of generations, cultural change can occur with extreme rapidity and can be measured in terms of years. This raises the possibility that cultural change may move so fast that it exceeds the biological capacities of man in respect to social behaviour.

Here again we are handicapped in our scientific information. There is no way of restoring precul-tural man to life and discovering what sort of social existence he led. Judging from existing animal societies, however, we may assume that his social behaviour was well adapted to a particular kind of stable social organisation. Beyond this, all we can do is to observe present-day human beings and study their social behaviour under varying conditions. We know that sexual behaviour plays an important role. Instead of definitely limited periods of sexual behaviour, human females have very nearly reached the situation of constant receptivity, if not constant estrus, and there is no seasonal limitation on either sex.

We know also that, because of the relatively slow developmental rate of human infants, care-giving behaviour is highly important, as is the care-soliciting behaviour of the young. It is likewise obvious that care-giving behaviour is often extended to adults as well as to children and that it exists in both sexes. Young children are there-fore one of the focal points of human societies, and as with all primate societies, the relationship between old and young is an important one. As with any highly social species, we may assume that human beings have evolved behavioural capacities for the control of agonistic behaviour, partly through forming habits of not fighting while young and partly through the formation of dominance-subordination relationships. These facts are almost self-evident. It is not as well recognised that human beings are also highly allelomimetic, with strong tendencies to do what others around them are doing. Because of differences in age and sex, allelomimetic behaviour may be difficult in mixed groups, and we can observe a tendency for human groups to sort themselves out on the basis of age and sex.

Besides these major types of advanced social behaviour, human beings continue to show all of the more primitive sorts of social behaviour, for example, ingestive, eliminative, shelter-seeking, and investigative. In regard to the last, another characteristic of human beings is that they are highly curious. We can conclude that human beings show all of the basic kinds of social behaviour in strongly developed form, and we can advance the hypothesis that any successful form of cultural organisation of society must provide for reasonably satisfactory development and expression of these types of behaviour. Conversely, any form of organisation which attempts to completely suppress or distort a fundamental kind of behaviour will result either in the dissolution of the society or maladaptive behaviour on the part of the individual within it.

TYPES OF BEHAVIOR

Observations on sexual behavior of female farm animals are useful in implementing breeding programs. Males and females of certain species produce *pheromones,* chemical substances that attract the opposite sex. Cows, ewes, and mares may have pheromones present in vaginal secretions and urine when they are in heat. Bulls, rams, and stallions will smell these pheromones. A common behavioral response in this process is called flehmen, during which the male animals lifts its head and curls its upper lip.

When females are sexually receptive, they usually seek out a male if mating has not previously occurred. Females are receptive for varying lengths of time; cows are in heat for approximately 12 hours, whereas mares show heat for 5-7 days with ovulation occurring during the last 24 hours of estrus.

It appears that in a sexually active group of cows, the bull is attracted to a cow in heat most often by visual means (observing cow-to-cow mounting) rather than olfactory clues. Cows that are in heat, for example, allow themselves to be mounted by others. Producers observe this condition of "standing heat" or estrus.

The ram chases a ewe that is coming into heat. The ram champs and licks, puts his head on the side of the ewe, and strikes with his foot. When the ewe reaches standing heat, she stands when approached by the ram.

The buck goat snorts when he detects a doe in heat. The doe shows unrest and may be fought by other does. Mating in goats is similar to that in sheep.

The boar does not seem to detect a sow that is in heat by smelling or seeing. If introduced into a group of sows, a boar chases any sow in the group. The sow that is in heat seeks out the boar for mating, and when the boar is located she stands still and flicks her ears. Boars produce pheromones in the saliva and preputial pouch, which attracts sows and gilts in estrus to the boars.

The sequence of events in estrus detection in horses appears to be similar to swine (previously described). The stallion approaches a mare from the front and a mare not in heat runs and kicks at the stallion. When the mare is in standing heat she stands, squats somewhat, and urinates as he approaches. Her vulva exhibits winking (opens and closes) when she is in heat.

Male chickens and turkeys show a preference for certain females and may even refuse to mate with other females. Likewise, the females may refuse to mate with certain males. This is a serious problem when pen mating of one male and 10-15 females are practiced. The eggs of some females may be infertile. The preferential mating is a greater problem in chickens, as AI is the common breeding practice in turkeys.

Some profound behavior patterns are associated with sex of the animal and changes resulting from castration. This verifies the importance of hormonal-directed expression of behavior. Intact males have more aggressive behavior, whereas castrates are more docile after losing their source of male hormone.

Taken from Biology, The Dynamics of Life

Much of an animal's *courtship behavior* is instinctive. Courtship behavior is a behavior that males and females of a species carry out before mating. Like other instinctive behaviors, courtship has evolved through natural selection. Imagine what would happen to the survival of a species if members were unable to recognize other members of the same species.

Different species of fireflies, for example, can be seen at dusk flashing distinct light patterns. However, female fireflies of one species respond only to those males exhibiting the species-correct flashing pattern.

Some courtship behaviors help prevent females from killing males before they have had the opportunity to mate. For example, in some spiders, the male is smaller than the female and risks the chance of being eaten if he approaches her. Before mating, the male presents the female with a nuptial gift, an insect wrapped in a silk web. Wile the female is unwrapping and eating the insect, the male is able to mate with her without being attacked. After mating, however, the male may be eaten by the female anyway.

FIXED ACTION PATTERNS (FAP)

The term Fixed Action Pattern (FAP), or Modal Action Pattern (MAP), is sometimes used in ethology to denote an instinctive behavioral sequence that is relatively invariant within the species and almost inevitably runs to completion.

Fixed action patterns, or similar behaviour sequences, are produced by a neural network known as the innate releasing mechanism in response to an external sensory stimulus known as a sign stimulus or releaser. A fixed action pattern is one of the few types of behaviors which is thought to be "hard-wired" and instinctive.

Konrad Lorenz was one the key founders of the fixed action pattern definition. He identified six characteristics of fixed action patterns. That stated fixed action patterns are, stereotyped, complex, species characteristic, released, triggered and, independent of experience.

Terminology

The term "sign stimulus", or "releaser", is used to denote a simple feature of a complex stimulus that can elicit a FAP. For example, the red belly of a male stickleback elicits a head-down, attack behaviour in other male sticklebacks.

This same response can be elicited by artificial models or objects that contain the sign stimulus of red, for example, a red coloured card. The terms "sign stimulus" and "releaser" are sometimes used interchangeably; however, they have different meanings. The term "sign stimulus" is used to denote a feature of an animal's environment that elicits a particular response. The term "releaser" is used for a stimulus that has evolved to facilitate communication between conspecifics (animals of the same species).

Examples

Male stickleback behaviour

One example of fixed action patterns is the courtship and aggression behaviour of the male three-spined stickleback during the mating season, described in a series of studies by Niko Tinbergen.

During spring, male sticklebacks change colour, establish a territory and build a nest. They attack male sticklebacks that enter their territory, but court females and entice them to enter the nest to lay eggs.

Tinbergen used crude models of sticklebacks to investigate which features of male and female sticklebacks elicited attack and courtship behaviour from male sticklebacks. Tinbergen's main findings were that male sticklebacks responded in a relatively invariant way and attacked a model with a red belly, but in contrast, courted a model with a swollen belly.

Egg-retrieval behaviour

Another example of a behaviour that has been described as a FAP is the egg-retrieval behavior of the greylag goose, reported in classic studies by Niko Tinbergen and Konrad Lorenz. Like many ground-nesting birds, if an egg becomes displaced from the nest, the greylag rolls it back to the nest with its beak. The sight of the displaced egg is the sign stimulus and elicits the egg-retrieval behaviour. If the egg is removed from the goose during the performance of egg-rolling, the bird often continues with the behavior, pulling its head back as if an imaginary egg is still being maneuvered by the underside of its beak. The greylag will also attempt to retrieve other egg-shaped objects, such as a golf ball, door knob, or even a model egg too large to have possibly been laid by the goose itself (i.e. a supernormal stimulus).

Others

Fig. : Kelp gull chicks peck at red spot on mother's beak to stimulate regurgitating reflex.

Other examples include:

- Some mating dances, commonly carried out by birds, are examples of fixed action patterns. In these cases, the sign stimulus is typically the presence of the opposite sex.
- Kelp gull chicks are stimulated by a red spot on the mother's beak to peck at the spot (pictured), which induces regurgitation.
- Some moths instantly fold their wings and drop to the ground if they encounter ultrasonic signals such as those produced by bats; see ultrasound avoidance.
- Mayflies drop their eggs when they encounter a certain pattern of light polarization which indicates they are over water.
- Some lizards and snakes exhibit strike-induced chemosensory searching behavior as a means of relocating bitten and envenomated prey by scent-trailing.

Significance

FAPs, or inflexible behavior patterns, are usually noticeable in animal behavior because FAPs represent the simplest type of behavior in which a particular *stimulus* almost always results in an invariable behavioral response.

FAPs have been said to be "hard-wired" in animals' brains. They are unusual in that they are relatively *un-influenced* by the environment, once the behaviour has been elicited. The existence of FAPs is rather unusual in that a fixed response can lead to maladaptive results, whereas flexible behaviors are generally more likely to be adaptive by increasing fitness.

Because of this, most behaviors which are both FAPs and occur in more complex animals are usually essential to the animal's fitness, or in which speed (i.e. an absence of learning) is a factor. For instance, the greylag goose's egg rolling behavior is so essential to the survival of its chicks that the fitness of the parent bird is increased by the behavior being relatively invariant. A stickleback will attack any male fish who enters his territory while the female is sexually receptive, reacting to their red colour, while the female stickleback triggers behavior in the male resulting in the fertilization of her eggs.Relatively invariant behaviors are also predictable, which can lead to their exploitation by humans or other animals.

There are exceptions to fixed action pattern rules. For example, male blond ring doves isolated from their own species will resort to courting a pigeon, then a human's hand, and finally expressing courtship activity alone in their box if they are left alone for a long period of time.

2

Developmental Biology

Recent findings in Developmental Biology concerning the signaling pathways that lead to the production of many different cell types during embryonic stages of life are vital to, and advance our knowledge of, the normal processes involved in tissue maintenance and cell replacement during adult life. Such signaling pathways are the basis of the "turning on and off" of specific panels of genes. These genes are controlled by a limited number of regulatory genes that react to the signals, thereby setting up a "molecular memory" to the cell and giving it a specific identity.

In the context of multifactorial disease, trauma and the environment, innovative genetic and biochemical approaches in Developmental and Stem Cell Biology offer great potential for clinical intervention in the form of novel molecular and cellular therapies. This innovative knowledge biomedical technology research programme concerns Stem Cells in Development and Disease and aims to identify and characterize the genetic cascades and regulatory pathways that control cell identity throughout development in several stem cell and tissue systems.

To achieve these goals, we will utilize integrated biochemical and functional genetic approaches, combining gene expression analysis, novel gene disruption technology, stem cell culture systems and rapid monoclonal antibody production. Bioinformatics databases will be integrated using the latest three-dimensional visualization technology and will lead to a new perception of multifactorial processes. This in turn will lead to validation and the subsequent formulation of novel intervention strategies for disease and trauma.

The knowledge programme, Stem Cells in Development and Disease, integrates the knowledge infrastructure of 11 internationally competitive Dutch research laboratories specializing in developmental, stem cell and molecular biology, together with 4 biotechnology companies, to advance biomedical technology and applications in human health and disease. The recent high interest in cell replacement therapies for multifactorial diseases, trauma and environmental circumstances relies to a large extent on insights coming out of the field of Developmental Biology. Just last summer, a group of Developmental Biologists discovered the complete series of inductive signals and transcription factors necessary to recapitulate the normal generation of motor neurons (a finding of great interest to spinal cord injury patients).

These fully functional motor neurons were generated from embryonic stem (ES) cells, the most primitive culturable source of cells for all cell lineages. Thus, inductive signals and transcription factors involved in the normal developmental pathways during embryogenesis can be used in the future to direct ES cells or other somatic cells to form specific classes of cells to be used in vivo. The Netherlands has been on the forefront of Developmental Biology, beginning with the findings of Nieuwkoop and most recently, with important contributions in the understanding of signaling pathways. For example, the groups of Clevers (Utrecht) and Fodde (Leiden) have shown that the Wnt developmental signaling pathway affects the differentiation of gut stem cells and thus is implicated in intestinal-related disease and tumorigenesis.

Using such developmental strategies, the outcome of these findings and the many other findings of the Dutch Developmental Biology community most certainly will lead to future clinical therapeutic applications for developmental defects, trauma and tumorigenesis. From the earliest stages of embryogenesis, and particularly during the stages of cell fate determination, organogenesis and homeostasis/normal cell replacement, defective developmental processes result in a wide range of syndromes and diseases. These include for example thalassemia, sickle cell anemia, and particularly childhood leukemias in the hematopoietic system; haemorrhagic telangiectasia in the circulatory system; multiple sclerosis and Charcot Marie Tooth disease in the nervous system; hypospadias and infertility syndromes in the genital and gonadal system; craniofacial malformations and congenital malformations of limbs.

By understanding the genetic programmes and signaling pathways leading to normal development, we will be better able to control the environmental and genetic factors that lead to infertility, abnormal fetal growth and defective formation and functioning of complex organ systems. The basis of development is within the genes. Indeed, the genetic programme leading to production of all the different cell types to form an embryo and ultimately the adult, begins shortly after fertilization. As the fertilized egg begins to divide, the complex regulation of gene expression results in changes within the exponentially increasing number of embryonic cells and in their interactions with each other, leading to the specification of cells for all the organ systems.

The understanding of genetic programmes and the cell-cell interactions that influence these programmes particularly in stem cells, is by no means complete. New genes and signaling pathways have yet to be identified. Complete signaling pathways in cell fate determination and differentiation need to be elucidated. Underlying principles controlling stem cell growth and differentiation are yet to be uncovered. Novel genetic and biochemical approaches offer enormous opportunities to extend our knowledge of normal development and stem cell biology, through not only the identification and classification of key regulatory molecules and signaling cascades, but also through functional testing regimes. Ultimately, these findings should lead to advances in disease treatments through cell replacement therapies and molecular intervention.

The aims of this knowledge biomedical technology research project, Stem Cells in Development and Disease, are to identify and characterize the signaling cascades and regulatory pathways that control stem cell differentiation throughout development. This will be achieved through integrated functional genetic approaches and databases, combining a novel gene targeting ES cell strategy, Cre-lox technology, 3D optical microscopy and tomography on embryos and tissues, gene and protein expression analysis, stem cell cultures and rapid single chain monoclonal antibody generation. The eludication and understanding of developmental signaling and regulatory processes will be of great relevance for future therapeutic intervention both at the molecular and cellular levels.

WHAT IS BIOLOGY?

Biology is the study of life. In studying biology, you will frequently run across a lot of "big words" that you will need to learn. One thing that can help you understand those words, which will also aid you in everything from reading the newspaper to communicating with your doctor, is to know the Greek or Latin (or other) derivations of the wordstems which make up those words. For example, the word "biology" is made from the wordstems bios, which means "life," and -logy which means "to study" or "the study of." Is biology important in your everyday life? How? What sorts of questions about biology are or should be important to non-majors, to general people? What branches or subdisciplines within biology might have the answers to those questions? Can you think of any other branches of biology and what they study?

The word biology is derived from the greek words/bios/meaning/life/ and/logos/meaning/study/and is defined as the science of life and living organisms. An organism is a living entity consisting of one cell e.g. bacteria, or several cells e.g. animals, plants and fungi. Aspects of biological science range from the study of molecular mechanisms in cells, to the classification and behaviour of organisms, how species evolve and interaction between ecosystems.

Biology is the study of living organisms. It is concerned with the characteristics, classification, and behaviours of organisms, how species come into existence, and the interactions they have with each other and with the natural environment. Biology encompasses a broad spectrum of academic fields that are often viewed as independent disciplines. However, together they address phenomena related to living organisms (biological phenomena) over a wide range of scales, from biophysics to ecology. All concepts in biology are subject to the same laws that other branches of science obey, such as the laws of thermodynamics and conservation of energy.

Biology encompasses a broad spectrum of academic fields that are often viewed as independent disciplines. However, together they address phenomena related to living organisms (biological phenomena) over a wide range of scales, from biophysics to ecology. It is concerned with the characteristics, classification, and behaviours of organisms, how species come into existence, and the interactions they have with each other and with the natural environment. There are however unifying features. All biological phenomena are subject to the same laws that other branches of science obey, such as the laws of thermodynamics and conservation of mass. Practically the continued existence of any biological organism is always dependant on acquiring sufficient energy to overcome deleterious entropic effects.

Many of the subdisciplines of biology, such as botany, zoology, and medicine are ancient. However, biology as a unified science first developed in the nineteenth century, as scientists discovered that all living things shared certain fundamental characteristics and were best studied as a whole. Today biology is one of the most prominent scientific fields. Over a million papers are published annually in a wide array of biology and medicine journals, and biology is a standard subject of instruction at schools and universities around the world.

As such a vast field, biology is divided into a number of subdisciplines. The old division by type of organism remains with botany encompassing the study of plants, zoology the study of animals, microbiology the study of microorganisms, and other similar disciplines. The field is also divided by the scale being studied: molecular biology looks at the fundamental chemistry of life; cellular biology looks a the basic building block of all life the cell; Physiology looks at the internal structure of organism; and ecology looks at how various organisms interrelate.

Biology often overlaps with other sciences; for example, biochemistry and toxicology with biology, chemistry, and medicine; biophysics with biology and physics; stratigraphy with biology and geography; astrobiology with biology and astronomy. Social sciences such as geography, philosophy, psychology and sociology can also interact with biology, for example, in administration of biological resources, developmental biology, biogeography, evolutionary psychology and ethics.

Biology is the study of living organisms and as such, it's one of the broadest subject areas. It ranges from the sub-cellular to whole organisms and environments. At Newcastle we concentrate on studying animals, plants and micro-organisms and give you the chance to maintain breadth throughout your degree. Biological knowledge is increasing at an unprecedented rate. New techniques such as genetic modification and genome sequencing give us extraordinary new tools for understanding how life works and for using that understanding - but with our new abilities also come new responsibilities.

Biology helps us:

- Understand ourselves and the world around us;
- Increase the quality of human life - better use of natural resources, preventing diseases and developing new ways to use living organisms, for example;
- Improve the effects humans have on our environment - such as conserving natural habitats and endangered species or reducing the effects of environmental pollution;
- Make judgements on how we should use our knowledge.

Biology is more important than ever in today's world! Biology is the study of life in all its various and rich forms. It influences our everyday experiences, from our relationships with the environment, to the foods we consume, to the way we treat diseases. In fact, functioning in the present world without biological knowledge is a distinct handicap, since biological advances are at the forefront of social change.

HISTORY OF BIOLOGY

The history of biology traces human understanding of the living world from the earliest recorded history to modern times. Though the concept of biology as a single coherent field of knowledge only arose in the 19th century, the biological sciences emerged from traditions of medicine and natural history reaching back to the ancient Greeks, particularly Galen and Aristotle, respectively.

Anaximander, a Greek philosopher who lived from 611 to 546 BC, is credited with the first written work on natural science, a classical poem entitled On Nature. In this poem, he presented what may be the first written theory of evolution He said that in the beginning there was a fish-like creature with scales etc. that arose in and lived in the world ocean. As some of these advanced, they moved onto land, shed their scaly coverings, and became the first humans.

In 570 BC, Xenophanes was born. He was one of the first people to write about his observations of fossils. He thought that fossils were an indication that there was water/mud previously in an area.

Hippocrates lived from about 400 to 300 BC. One of the things for which he is remembered is his theory that the human body was composed of the four elements (earth, air, fire, water) plus four fluids or humors: sanguis or blood, produced by the heart; choler or yellow bile, produced by the liver; melancholia or black bile, produced by the spleen; and phlegma or phlegm, produced by the brain, which corresponded with these.

Galen lived from 131 to 210 AD. He was a philosopher, physician, anatomist, and is famous for his descriptions of human anatomy which were considered authoritative for the next 1000 years. He associated the four humors with Empedocles' scheme (blood or sanguis, yellow bile or choler, black bile or melancholia, and phlegma or phlegm. He and subsequent writers worked out and added to an elaborate system that included the four organs from which the four humors came, the four seasons, the four stages in human life, and several other things that came in fours.

In the Twelfth to Fourteenth Centuries, in Europe, science, culture, etc. were at their lowest, hence this time is referred to as the Dark Ages. Much of the knowledge from the Greeks was lost in European civilization, but preserved among the Arabs.

In 1665, Robert Hooke was the first person to see and name cells. He examined (dead) cork bark with a primitive microscope and saw little cubicles which he called cells (cell = room, cubicle).

Anton van Leeuwenhoek was the first person to observe sperm cells and with his very primitive microscope, thought he saw tiny body parts in the sperm. He used this as "proof" of the idea that the homunculus was in the sperm and the mother's body just served as a place for the planted seed to grow. Apparently Leeuwenhoek, himself, did not claim the sperm was the homunculus, but that sperm provided major life-giving qualities. Regnier de Graaf, on the other hand, thought he saw the homunculus in the egg and that the sperm merely triggers its growth somehow. This split the preformationists: was the miniature adult in the egg (followers of de Graaf) or the sperm (followers of Leeuwenhoek)? Additionally, Leeuwenhoek proposed that fertilization occurs when the sperm enters the egg, but this could not actually be observed for another 100 years because of the quality of microscopes which were available. Most of the subsequent arguments as to the nature of embryos were based on speculation rather than research.

In 1668, Francesco Redi did an experiment with flies and open vs. closed flasks which contained meat. His hypothesis was that rotten meat does not turn into flies. He observed these flasks to see in which one(s) maggots developed. He found that if a flask was closed with a lid so adult flies could not get in, no maggots developed on the rotting meat within. In a flask without a lid, maggots soon were seen in the meat, followed by more adult flies. The results of this experiment disproved the idea of spontaneous generation for larger organisms, but people still thought microscopic organisms like algae or bacteria could arise that way.

Karl von Linné lived from 1707 to 1778. As was the custom among scholars of that time, he Latinized his name, which became Carolus Linnaeus. He gets credit for our present-day classification scheme and the system of two-part scientific names for organisms, thus has been given the nickname, "The Father of Taxonomy." He was the first to attempt to classify organisms for their own sake (based on things like similar body structures) rather than to serve some human use.

Linnaeus believed that people were supposed to try to discover and decipher the Creator's scheme, which would become evident as organisms were grouped with "relatives." He gave two-part Latin names to each organism he knew. For example: Homo sapiens is the scientific name for humans. Note that the genus name (the first name) is always capitalized, but the species name (the second one) never is.

Because this is in a foreign language (Latin), when using a computer (or having something done by a printer), it is proper to italicize scientific names. To indicate this when writing or typing a scientific name, you should underline the name. Also, note the spelling of this particular scientific name. The species name, sapiens, is one of those words that ends in "s" (like scissors), and is both singular and plural. Another thing Linnaeus did was an attempt to organize all known organisms into a taxonomic hierarchy which he invented. The levels in this hierarchy, in order are:

James Hutton lived from 1726 to 1797. In 1795, he published a paper, later refined/modified by Charles Lyell (1797-1875), in which he said that land forms can be accounted for by current mechanisms; for example, a gorge was cut by the river running through it, and was not always there. From this, he drew two conclusions:

In 1798, Rev. Thomas Malthus, a sociologist, published a paper in which he looked at conditions in the poor neighbourhoods of London. He said that in humans, the problems of disease, suffering, starvation etc. were a consequence of the potential for the human population to grow faster than technology could keep up with. Things like the supplies of food, medical care, etc. were limited in comparison to the size of the population, thus there was competition for available resources and only the strong and healthy would survive. He was, thus, the first to talk about survival of the fittest.

In either 1802 or 1809, Jean Baptiste Lamarck published his theory of evolution. His main points were: evolution or change within a species is driven by an innate, inner striving toward greater perfection, use or disuse of various organs made them larger or smaller, accordingly, and these acquired traits could be inherited or passed on to offspring (inheritance of acquired traits).

A number of subsequent attempts were made to support or disprove this theory without the benefit of our modern knowledge of genetics. One experiment involved amputation of mouse tails for successive generations, showing that even after twenty generations, there was no effect: baby mice were still born with tails. The Jewish practice of circumcision was also cited as opposing evidence, since obviously it had caused no long-lasting change in the population and still needed to be done to each new boy baby. Lamarck's theory seemed to make sense in the light of the then-accepted theory of pangenes coming from the body parts to make up the homunculus. The classic example he used was giraffes. He felt that giraffes' necks got longer because they stretched to reach higher leaves, and this was passed on to their babies. Another example, to make the fallacy of his theory more apparent, would be two people who developed large arm muscles because they were blacksmiths, tennis players, or weight-lifters having a baby who was born with larger than normal arm muscles.

In 1828, Karl von Baer published on the developmental stages in mammalian eggs. He was able to show that an undifferentiated, single-celled egg grows into a many-celled embryo in which the cells all have different functions. This disproved the preformation theory (which said that the preformed homunculus just gets bigger).

In 1859, Charles Darwin published The Origin of Species by Means of Natural Selection or the Preservation of Favoured Races in the Struggle for Life, more commonly known as The Origin of Species. In this landmark book, he made four main points:

a. Individuals, even siblings, in a population vary (there is variation),
b. These variations can be passed to offspring (are inherited—remember: he, too, thought this happened via pangenes),
c. (from Malthus) More offspring are produced than the environment can support, so there is competition for resources, and
d. Those individuals whose characteristics make them best suited to the environment live and reproduce and have more offspring (survival of the fittest).

In 1745 - 1748, John Needham, a Scottish clergyman and naturalist showed that microorganisms flourished in various soups that had been exposed to the air. He claimed that there was a "life force" present in the molecules of all inorganic matter, including air and the oxygen in it, that could cause spontaneous generation to occur, thus accounting for the presence of bacteria in his soups. He even briefly boiled some of his soup and poured it into "clean" flasks with cork lids, and microorganisms still grew there.

A few years later (1765 - 1767), Lazzaro Spallanzani, an Italian abbot and biologist, tried several variations on Needham's soup experiments. This initiated a heated argument between Needham and Spallanzani over sterilization (boiled broth in closed vs. open containers) as a way of refuting spontaneous generation. In 1864, Louis Pasteur claimed the prize money and published the results of an experiment he did to disproved spontaneous generation in these microscopic organisms. He used flasks of broth, as above, but did not seal any of them.

In 1865, Gregor Mendel, an Austrian monk, published a paper on genetics that earned him the nickname "the Father of Modern Genetics." One of Mendel's jobs at the monastery was to care for the garden. As he went about his chores, he noticed that some of his pea plants were tall while others were short, some had purple flowers while others had white, some had yellow seeds while others had green, and some had wrinkled seeds while others had smooth seeds.

As Mendel raised peas, he made specific crosses between certain plants and did something very unusual for biology in those days: he counted the results. From this he developed a theory of genetics that refuted the pangene/homunculus idea and enabled people to predict the outcome of a genetic cross if the genes of the parents were known. When Mendel first published his paper, the idea of the pangenes was still so deeply held that people ignored his work or dismissed it as false. It wasn't until 1900 that a couple of botanists working on other research rediscovered his work. We will discuss Mendel's theory in more depth when we talk about genetics.

In 1870 the process of mitosis, regular cell division by which one cell divides to make two cells, was observed, and researchers noticed that chromosomes, whose function was not understood, were moving around in the cell during mitosis so that each daughter cell got an exact set of them.

In 1890 the process of meiosis, a special cell division involved in producing eggs or sperm, was observed. Again, researchers did not yet understand what chromosomes were, but they did note that as a result of meiosis, each egg or sperm cell formed had half as many chromosomes as the original cell.

Subdisciplines of Biology:

Anatomy	*The study of body parts and theirana= up locations*	*tom = to cut*
Botany	The study of plants	botan = grass, pasture
Cytology	The study of cells	cyto = cell
Ecology	The study of the interrelations between organisms and their environment	eco = house
Entomology	The study of insects	entomol = insect
Genetics	The study of genes and heredity	gene = origin, birth
Microbiology	The study of bacteria and other microscopic organisms	micro = small
Molecular Biology	The study of the various molecules and chemical reactions that take place in organisms	molecul = a little mass
Paleontology	The study of formerly-living organisms like fossils and dinosaurs	paleo = ancient onto = being, existing
Physiology	The study of how various body parts function	physio = nature
Zoology	The study of animals	zoo = animal

In 1936, Alexander Ivanovich Oparin, a Russian scientist, published The Origins of Life, in which he described hypothetical conditions which he felt would have been necessary for life to first come into existence on early Earth. This included an atmosphere of methane, ammonia, and other gases, much volcanic activity, lightening, and warm soil and water temperatures. This hypothesis was later tested by an experiment done by Stanley Miller as a graduation student under Harold Urey in 1953. We will be discussing these ideas in more depth in Biology 106.

In 1953, James Watson, an American, and Francis Crick, an Englishman, published a paper in which they proposed a hypothetical structure for DNA which also showed how DNA could be the genetic code material and suggested a means whereby it could replicate itself. Subsequent chemical analyses of DNA have upheld their prediction.

Since that time, the Modern Synthetic Theory of Evolution was developed based on the theories of Darwin, Mendel, Malthus, Lyell, Oparin, and many others, including more recent work in genetics, geology, paleontology, and animal behaviour.

ORIGIN OF LIFE ON EARTH

So, what does all this have to do with the origin of life?

The genetic code represents an historical record of the development of the organism with an extraordinary amount of detail (825 megabytes is a lot of detail!). An organism which shares significant code sequences with another organism very likely has a common ancestor. By looking at changes in non-functional DNA we can estimate the time since that ancestor lived. By comparing genomes we can construct a "family tree" of life on Earth.

Based on data from the HGP and other sources we can say things like the following:

- All humans are descended from a single individual who lived about 270,000 years ago.
- Humans and New World monkeys share an ancestor which lived about 7 million years ago.
- Humans and mice share a common ancestor which lived about 50 million years ago.
- All life on earth is thought to be descended from an original primordial single cell organism which lived about 3.5 billion years ago.
- The Earth was formed about 4.5 billion years ago but was probably incompatible with life until perhaps 3.8 billion years ago so life apparently appeared relatively quickly.

As more genetic code data is available on various other organisms and as analysis of differences and similarities of codes progresses the entire family tree of life on earth will eventually be developed and more will be known about the characteristics of the primordial organism.

SO, where did that original primordial organism come from? There seem to be several schools of thought.

Theory 1 - Life Appeared Spontaneously

Some scientists believe that life arose spontaneously from available materials present on the early Earth. In fact, experiments have been conducted in which air, water, carbon dioxide, methane, and common minerals were "cooked" in the presence of energy sources such as heat, sunlight, and simulated lightning to see if life or precursors of life would appear. Indeed, organic "building blocks" such as amino acids did appear.

But it is a very, very long way from amino acids to a life form. The genetic work indicates that the complexity of genetic codes doesn't track that well with the apparent complexity of the organism and that even very simple organisms have quite complex genomes. The simplest known living thing is the microbe mycoplasma genitalium which causes human non-gonococcal urethritus. This microbe has a genetic code of about 570,000 base pairs. Viruses are simpler but aren't really "alive" in the sense that they cannot reproduce or grow without using the mechanisms in a living cell to do so. The bacteria e coli has a genetic code of about 5.7 million base pairs.

But e coli and mycoplasma can't live in the absence of other more complex organisms (e coli lives in animal gut, mycoplasma lives in ... well you get the idea). In fact the primordial organism must have been at the bottom of the food chain, capable of synthesizing its own food from non-living material, and living without assistance from any other living organism. It could have possibly been something on the order of blue-green algae which has 3.6 million base pairs in its genetic code and is thought to be about 3.5 billion years old. Mycoplasma, bacteria, and viruses all must have "devolved" from more complex organisms in response to the availability of more complex forms to act as hosts or links in the food chain.

The original organism had mechanisms (ability to grow, reproduce, and evolve) which led to the evolution of the diverse life forms which now exist on Earth and as indicated above this evolution is documented in the genetic codes of organisms now alive as well as in fossil evidence. But under this scenario the original organism would have had to appear by random happenstance aggregation of materials. This is somewhat like believing that because while digging you found a rock that looked like a brick, if you dug long and hard enough you would eventually find something that looked like the Sistine Chapel complete with Michelangelo's Creation on the ceiling.

Life Evolved from Simpler Organisms

Some scientists feel that life originated spontaneously as a much simpler organism than any now found. A difficulty with this idea is the absence of any current examples of the simpler organism. In general, appearance of more complex organisms has not resulted in disappearance of simpler forms. We still have cockroaches. We still have fruit flies. Indeed, there is evidence of devolution. Eventually, back tracking of many living genetic codes should enable some insight into the probability of this theory.

Its Unknowable

Some scientists take the view that the origin of the primordial organism is "unknowable" meaning that not only do we not know but we are unlikely to ever know and that the subject is therefore more appropriate for philosophy or religion than science. The origin of the primordial organism is therefore the biological equivalent of the "Big Bang Theory" in Astrophysics in which astrophysicists think the entire universe was once the size of a golf ball which then exploded to create the observed universe. They can trace observed cosmic phenomena such as galaxies, red shift, and background radiation back to the golf ball but they admit that it is "unknowable" as to how the golf ball got there.

It Came from Outer Space

Some believe that life originated elsewhere in the universe and was then somehow distributed. This doesn't have to mean biological contamination of the early Earth by space travelers flushing their ballast tanks. It could be that life was distributed via simple frozen or sporelated organisms carried by fragments of a destroyed planet or ejecta blasted into space by meteorite impact. Comets are known to contain water and ice and NASA thinks it has found evidence of fossilized bacteria in meteorites. DNA has been recovered from material 20 million years old.

The possibility that life originated somewhere else in the universe (it is a Very large universe) and then came here seems to many more likely than the idea that life originated on Earth. The space theory is also less egocentric. Keep in mind that all previous "Earth is the centre of the universe" theories have been disproved. A consequence of the space theory is that life might be widely distributed. Life might appear relatively rapidly on any planet that has appropriate conditions, at least in regions which were in a position to be seeded from the source – a sort of "universe as Petri dish" concept. In other words, if there is life on Earth, then there is likely to be life in any nearby system that has planets with appropriate conditions.

HISTORY OF THE CONCEPT OF LIFE IN SCIENCE

In a letter to Joseph Dalton Hooker of February 1 1871, Charles Darwin made the suggestion that the original spark of life may have begun in a "warm little pond, with all sorts of ammonia and phosphoric salts, lights, heat, electricity, etc. present, so that a protein compound was chemically formed ready to undergo still more complex changes". He went on to explain that "at the present day such matter would be instantly devoured or absorbed, which would not have been the case before living creatures were formed." In other words, the presence of life itself prevents the spontaneous generation of simple organic compounds from occurring on Earth today – a circumstance which makes the search for the origin of life dependent on the sterile conditions of the laboratory. An experimental approach to the question was beyond the scope of laboratory science in Darwin's day, and no real progress was made until 1924 when Aleksandr Ivanovich Oparin demonstrated that it was the presence of atmospheric oxygen and other more sophisticated life-forms that prevented the chain of events that would lead to the evolution of life.

In his The Origin of Life on Earth, Oparin argued that a "primeval soup" of organic molecules could be created in an oxygen-less atmosphere through the action of sunlight. These would combine in ever-more complex fashion until they dissolved into a coacervate droplet. These droplets would "grow" by fusion with other droplets, and "reproduce" through fission into daughter droplets, and so have a primitive metabolism in which those factors which promote "cell integrity" survive, those that don't become extinct. All modern theories of the origin of life take Oparin's ideas as a starting point.

Current Models on Origin of Life

There is no truly "standard" model of the origin of life. But most currently accepted models build in one way or another upon a number of discoveries about the origin of molecular and cellular components for life, which are listed in a rough order of postulated emergence: Plausible pre-biotic conditions result in the creation of certain basic small molecules (monomers) of life, such as amino acids. This was demonstrated in the Miller-Urey experiment by Stanley L. Miller and Harold C. Urey in 1953.

Phospholipids (of an appropriate length) can spontaneously form lipid bilayers, a basic component of the cell membrane.

The polymerization of nucleotides into random RNA molecules might have resulted in self-replicating ribozymes (RNA world hypothesis).

Selection pressures for catalytic efficiency and diversity result in ribozymes which catalyse peptidyl transfer (hence formation of small proteins), since oligopeptides complex with RNA to form better catalysts.

Thus the first ribosome is born, and protein synthesis becomes more prevalent.

Proteins outcompete ribozymes in catalytic ability, and therefore become the dominant biopolymer. Nucleic acids are restricted to predominantly genomic use.

The origin of the basic biomolecules, while not settled, is less controversial than the significance and order of steps 2 and 3. The basic chemicals from which life was thought to have formed arc methane (CH_4), ammonia (NH_3), water (H_2O), hydrogen sulfide (H_2S), carbon dioxide (CO_2) or carbon monoxide (CO), and phosphate (PO_4^{3-}). Molecular oxygen (O_2) and ozone (O_3) were either rare or absent.

As 2007, no one has yet synthesized a "protocell" using basic components which would have the necessary properties of life (the so-called "bottom-up-approach"). Without such a proof-of-principle, explanations have tended to be short on specifics.

However, some researchers are working in this field, notably Steen Rasmussen at Los Alamos National Laboratory and Jack Szostak at Harvard University. Others have argued that a "top-down approach" is more feasible. One such approach, attempted by Craig Venter and others at The Institute for Genomic Research, involves engineering existing prokaryotic cells with progressively fewer genes, attempting to discern at which point the most minimal requirements for life were reached. The biologist John Desmond Bernal, coined the term Biopoesis for this process, and suggested that there were a number of clearly defined "stages" that could be recognised in explaining the origin of life.

Stage 1: The origin of biological monomers

Stage 2: The origin of biological polymers

Stage 3: The evolution from molecules to cell

Bernal suggested that Darwinian evolution may have commenced early, some time between Stage 1 and 2.

ORIGIN OF ORGANIC MOLECULES

The Miller-Urey experiment attempted to recreate the chemical conditions of the primitive Earth in the laboratory, and synthesized some of the building blocks of life.

Miller's Experiments

In 1953 a graduate student, Stanley Miller, and his professor, Harold Urey, performed an experiment that proved organic molecules could have spontaneously formed on Early Earth from inorganic precursors. The now-famous "Miller-Urey experiment" used a highly reduced mixture of gases - methane, ammonia and hydrogen – to form basic organic monomers, such as amino acids. Whether the mixture of gases used in the Miller-Urey experiment truly reflects the atmospheric content of Early Earth is a controversial topic. Other less reducing gases produce a lower yield and variety. It was once thought that appreciable amounts of molecular oxygen were present in the prebiotic atmosphere, which would have essentially prevented the formation of organic molecules; however, the current scientific consensus is that such was not the case.

A recent experiment showed that a thick organic haze might have blanketed Early Earth. An organic haze can form over a wide range of methane and carbon dioxide concentrations, believed to be present in the atmosphere of Early Earth. After forming, these organic molecules would have floated down all over the Earth, allowing life to flourish globally.

Simple organic molecules are of course a long way from a fully functional self-replicating life form. But in an environment with no pre-existing life these molecules may have accumulated and provided a rich environment for chemical evolution ("soup theory"). On the other hand, the spontaneous formation of complex polymers from abiotically generated monomers under these conditions is not at all a straightforward process. Besides the necessary basic organic monomers, also compounds that would have prohibited the formation of polymers were formed in high concentration during the experiments.

Other sources of complex molecules have been postulated, including sources of extra-terrestrial stellar or interstellar origin. For example, from spectral analyses, organic molecules are known to be present in comets and meteorites. In 2004, a team detected traces of polycyclic aromatic hydrocarbons (PAH's) in a nebula, the most complex molecule, to that date, found in space. The use of PAH's has also been proposed as a precursor to the RNA world in the PAH world hypothesis.

It can be argued that the most crucial challenge unanswered by this theory is how the relatively simple organic building blocks polymerise and form more complex structures, interacting in consistent ways to form a protocell. For example, in an aqueous environment hydrolysis of oligomers/polymers into their constituent monomers would be favored over the condensation of individual monomers into polymers. Also, the Miller experiment produces many substances that would undergo cross-reactions with the amino acids or terminate the peptide chain.

Eigen's Hypothesis

In the early 1970s a major attack on the problem of the origin of life was organised by a team of scientists gathered around Manfred Eigen of the Max Planck Institute. They tried to examine the transient stages between the molecular chaos in a prebiotic soup and the transient stages of a self replicating hypercycle, between the molecular chaos in a prebiotic soup and simple macromolecular self-reproducing systems.

In a hypercycle, the information storing system (possibly RNA) produces an enzyme, which catalyzes the formation of another information system, in sequence until the product of the last aids in the formation of the first information system. Mathematically treated, hypercycles could create quasispecies, which through natural selection entered into a form of Darwinian evolution. A boost to hypercycle theory was the discovery that RNA, in certain circumstances forms itself into ribozymes, a form of RNA enzyme.

Wächtershäuser's Hypothesis

Another possible answer to this polymerization conundrum was provided in 1980s by Günter Wächtershäuser, in his iron-sulphur world theory. In this theory, he postulated the evolution of (bio)chemical pathways as fundamentals of the evolution of life. Moreover, he presented a consistent system of tracing today's biochemistry back to ancestral reactions that provide alternative pathways to the synthesis of organic building blocks from simple gaseous compounds.

In contrast to the classical Miller experiments, which depend on external sources of energy (such as simulated lightning or UV irradiation), "Wächtershäuser systems" come with a built-in source of energy, sulfides of iron and other minerals (e.g. pyrite). The energy released from redox reactions of these metal sulfides is not only available for the synthesis of organic molecules, but also for the formation of oligomers and polymers. It is therefore hypothesized that such systems may be able to evolve into autocatalytic sets of self-replicating, metabolically active entities that would predate the life forms known today.

The experiment as performed, produced a relatively small yield of dipeptides (0.4% to 12.4%) and a smaller yield of tripeptides (0.003%) and the authors note that: "under these same conditions dipeptides hydrolysed rapidly." Another criticism of the result is that the experiment did not include any organomolecules that would most likely cross-react or chain-terminate.

The latest modification of the iron-sulphur-hypothesis was provided by William Martin and Michael Russell in 2002. According to their scenario, the first cellular life forms may have evolved inside so-called black smokers at seafloor spreading zones in the deep sea. These structures consist of microscale caverns that are coated by thin membraneous metal sulfide walls. Therefore, these structures would solve several critical points of the "pure" Wächtershäuser systems at once:

The micro-caverns provide a means of concentrating newly synthesised molecules, thereby increasing the chance of forming oligomers; The steep temperature gradients inside a black smoker allow for establishing "optimum zones" of partial reactions in different regions of the black smoker (e.g. monomer synthesis in the hotter, oligomerisation in the colder parts); the flow of hydrothermal water through the structure provides a constant source of building blocks and energy (freshly precipitated metal sulfides);

the model allows for a succession of different steps of cellular evolution (prebiotic chemistry, monomer and oligomer synthesis, peptide and protein synthesis, RNA world, ribonucleoprotein assembly and DNA world) in a single structure, facilitating exchange between all developmental stages; synthesis of lipids as a means of "closing" the cells against the environment is not necessary, until basically all cellular functions are developed.

This model locates the "last universal common ancestor" (LUCA) inside a black smoker, rather than assuming the existence of a free-living form of LUCA. The last evolutionary step would be the synthesis of a lipid membrane that finally allows the organisms to leave the microcavern system of the black smokers and start their independent lives. This postulated late acquisition of lipids is consistent with the presence of completely different types of membrane lipids in archaebacteria and eubacteria (plus eukaryotes) with highly similar cellular physiology of all life forms in most other aspects.

Another unsolved issue in chemical evolution is the origin of homochirality, i.e. all monomers having the same "handedness" (amino acids being left handed, and nucleic acid sugars being right handed). Homochirality is essential for the formation of functional ribozymes (and probably proteins too). The origin of homochirality might simply be explained by an initial asymmetry by chance followed by common descent. Work performed in 2003 by scientists at Purdue identified the amino acid serine as being a probable root cause of organic molecules' homochirality. Serine forms particularly strong bonds with amino acids of the same chirality, resulting in a cluster of eight molecules that must be all right-handed or left-handed. This property stands in contrast with other amino acids which are able to form weak bonds with amino acids of opposite chirality. Although the mystery of why left-handed serine became dominant is still unsolved, this result suggests an answer to the question of chiral transmission: how organic molecules of one chirality maintain dominance once asymmetry is established.

"GENES FIRST" MODELS: THE RNA WORLD

The RNA world hypothesis suggests that relatively short RNA molecules could have spontaneously formed that were capable of catalyzing their own continuing replication. It is difficult to gauge the probability of this formation. A number of theories of modes of formation have been put forward. Early cell membranes could have formed spontaneously from proteinoids, protein-like molecules that are produced when amino acid solutions are heated - when present at the correct concentration in aqueous solution, these form microspheres which are observed to behave similarly to membrane-enclosed compartments. Other possibilities include systems of chemical reactions taking place within clay substrates or on the surface of pyrite rocks. Factors supportive of an important role for RNA in early life include its ability to replicate; its ability to act both to store information and catalyse chemical reactions (as a ribozyme); its many important roles as an intermediate in the expression and maintenance of the genetic information (in the form of DNA) in modern organisms; and the ease of chemical synthesis of at least the components of the molecule under conditions approximating the early Earth.

A number of problems with the RNA world hypothesis remain, particularly the instability of RNA when exposed to ultraviolet light, the difficulty of activating and ligating nucleotides and the lack of available phosphate in solution required to constitute the backbone, and the instability of the base cytosine (which is prone to hydrolysis). Recent experiments also suggest that the original estimates of the size of an RNA molecule capable of self-replication were most probably vast underestimates. More-modern forms of the RNA World theory propose that a simpler molecule was capable of self-replication (that other "World" then evolved over time to produce the RNA World). At this time however, the various hypotheses have incomplete evidence supporting them. Many of them can be simulated and tested in the lab, but a lack of undisturbed sedimentary rock from that early in Earth's history leaves few opportunities to test this hypothesis robustly.

"METABOLISM FIRST" MODELS: IRON-SULPHUR WORLD AND OTHERS

Several models reject the idea of the self-replication of a "naked-gene" and postulate the emergence of a primitive metabolism which could provide an environment for the later emergence of RNA replication.

One of the earliest incarnations of this idea was put forward in 1924 with Alexander Oparin's notion of primitive self-replicating vesicles which predated the discovery of the structure of DNA. More recent variants in the 1980s and 1990s include Günter Wächtershäuser's iron-sulphur world theory and models introduced by Christian de Duve based on the chemistry of thioesters. More abstract and theoretical arguments for the plausibility of the emergence of metabolism without the presence of genes include a mathematical model introduced by Freeman Dyson in the early 1980s and Stuart Kauffman's notion of collectively autocatalytic sets, discussed later in that decade.

However, the idea that a closed metabolic cycle, such as the reductive citric acid cycle, could form spontaneously (proposed by Günter Wächtershäuser) remains unsupported. According to Leslie Orgel, a leader in origin-of-life studies for the past several decades, there is reason to believe the assertion will remain so. Orgel summarizes his analysis of the proposal by stating, "There is at present no reason to expect that multistep cycles such as the reductive citric acid cycle will self-organize on the surface of FeS/FeS2 or some other mineral." It is possible that another type of metabolic pathway was used at the beginning of life. For example, instead of the reductive citric acid cycle, the "open" acetyl-CoA pathway (another one of the four recognised ways of carbon dioxide fixation in nature today) would be even more compatible with the idea of self-organisation on a metal sulfide surface. The key enzyme of this pathway, carbon monoxide dehydrogenase/acetyl-CoA synthase harbours mixed nickel-iron-sulphur clusters in its reaction centres and catalyses the formation of acetyl-CoA (which may be regarded as a modern form of acetyl-thiol) in a single step.

BUBBLE THEORY

Waves breaking on the shore create a delicate foam composed of bubbles. Winds sweeping across the ocean have a tendency to drive things to shore, much like driftwood collecting on the beach. It is possible that organic molecules were concentrated on the shorelines in much the same way. Shallow coastal waters also tend to be warmer, further concentrating the molecules through evapouration. While bubbles comprised of mostly water burst quickly, oily bubbles happen to be much more stable, lending more time to the particular bubble to perform these crucial experiments.

The phospholipid is a good example of an oily compound believed to have been prevalent in the prebiotic seas. Because phospholipids contain a hydrophilic head on one end, and a hydrophobic tail on the other, they have the tendency to spontaneously form lipid membranes in water. A lipid monolayer bubble can only contain oil, and is therefore not conducive to harbouring water-soluble organic molecules. On the other hand, a lipid bilayer bubble can contain water, and was a likely precursor to the modern cell membrane. If a protein came along that increased the integrity of its parent bubble, then that bubble had an advantage, and was placed at the top of the natural selection waiting list. Primitive reproduction can be envisioned when the bubbles burst, releasing the results of the experiment into the surrounding medium. Once enough of the 'right stuff' was released into the medium, the development of the first prokaryotes, eukaryotes, and multicellular organisms could be achieved. This theory is expanded upon in the book, "The Cell: Evolution of the First Organism" by Joseph Panno.

Similarly, bubbles formed entirely out of protein-like molecules, called microspheres, will form spontaneously under the right conditions. But they are not a likely precursor to the modern cell membrane, as cell membranes are composed primarily of lipid compounds rather than amino-acid compounds (for types of membrane spheres associated with abiogenesis,see protobionts, micelle, coacervate).

Hybrid Models

A growing realization of the inadequacy of either pure "genes-first" or "metabolism-first" models is leading the trend towards models that incorporate aspects of each.

OTHER MODELS

Autocatalysis

British ethologist Richard Dawkins wrote about autocatalysis as a potential explanation for the origin of life in his 2004 book. The Ancestor's Tale. Autocatalysts are substances which catalyze the production of themselves, and therefore have the property of being a simple molecular replicator. In his book, Dawkins cites experiments performed by Julius Rebek and his colleagues at the Scripps Research Institute in California in which they combined amino adenosine and pentafluorophenyl ester with the autocatalyst amino adenosine triacid ester (AATE). One system from the experiment contained variants of AATE which catalysed the synthesis of themselves. This experiment demonstrated the possibility that autocatalysts could exhibit competition within a population of entities with heredity, which could be interpreted as a rudimentary form of natural selection.

Clay Theory

A hypothesis for the origin of life based on clay was forwarded by Dr A. Graham Cairns-Smith of the University of Glasgow in 1985 and adopted as a plausible illustration by just a handful of other scientists (including Richard Dawkins). Clay theory postulates that complex organic molecules arose gradually on a pre-existing, non-organic replication platform — silicate crystals in solution. Complexity in companion molecules developed as a function of selection pressures on types of clay crystal is then exapted to serve the replication of organic molecules independently of their silicate "launch stage". It is, truly, "life from a rock."

Cairns-Smith is a staunch critic of other models of chemical evolution. However, he admits, that like many models of the origin of life, his own also has its shortcomings

Peggy Rigou of the National Institute of Agronomic Research (INRA), in Jouy-en-Josas, France reports that prions are capable of binding to clay particles and migrate off the particles when the clay becomes negatively charged. While no reference is made in the report to implications for origin-of-life theories, this research may suggest prions as a likely pathway to early reproducing molecules.

"Deep-hot Biosphere" Model of Gold

The discovery of nanobes (filamental structures smaller than bacteria containing DNA) in deep rocks, led to a controversial theory put forward by Thomas Gold in the 1990s that life first developed not on the surface of the Earth, but several kilometers below the surface. It is now known that microbial life is plentiful up to five kilometers below the earth's surface in the form of archaea, which are generally considered to have originated either before or around the same time as eubacteria, most of which live on the surface including the oceans. It is claimed that discovery of microbial life below the surface of another body in our solar system would lend significant credence to this theory. He also noted that a trickle of food from a deep, unreachable, source promotes survival because life arising in a puddle of organic material is likely to consume all of its food and become extinct.

"Primitive" Extraterrestrial Life

An alternative to Earthly abiogenesis is the hypothesis that primitive life may have originally formed extraterrestrially, either in space or on a nearby planet (Mars). (Note that exogenesis is related to, but not the same as, the notion of panspermia).

Organic compounds are relatively common in space, especially in the outer solar system where volatiles are not evapourated by solar heating. Comets are encrusted by outer layers of dark material, thought to be a tar-like substance composed of complex organic material formed from simple carbon compounds after reactions initiated mostly by irradiation by ultraviolet light. It is supposed that a rain of material from comets could have brought significant quantities of such complex organic molecules to Earth.

An alternative but related hypothesis, proposed to explain the presence of life on Earth so soon after the planet had cooled down, with apparently very little time for prebiotic evolution, is that life formed first on early Mars. Due to its smaller size Mars cooled before Earth(a difference of hundreds of millions of years), allowing prebiotic processes there while Earth was still too hot.

Life was then transported to the cooled Earth when crustal material was blasted off Mars by asteroid and comet impacts. Mars continued to cool faster and eventually became hostile to the continued evolution or even existence of life (it lost its atmosphere due to low volcanism), Earth is following the same fate as Mars, but at a slower rate.

Neither hypothesis actually answers the question of how life first originated, but merely shifts it to another planet or a comet. However, the advantage of an extraterrestrial origin of primitive life is that life is not required to have evolved on each planet it occurs on, but rather in a single location, and then spread about the galaxy to other star systems via cometary and/or meteorite impact. Evidence to support the plausibility of the concept is scant, but it finds support in recent study of Martian meteorites found in Antarctica and in studies of extremophile microbes. Additional support comes from a recent discovery of a bacterial ecosytem whose energy source is radioactivity.

EVOLUTION OF LIFE

Biologists studying evolution do a variety of things: population geneticists study the process as it is occurring; systematists seek to determine relationships between species and paleontologists seek to uncover details of the unfolding of life in the past. Discerning these details is often difficult, but hypotheses can be made and tested as new evidence comes to light. This section should be viewed as the best hypothesis scientists have as to the history of the planet. The material here ranges from some issues that are fairly certain to some topics that are nothing more than informed speculation. For some points there are opposing hypotheses — I have tried to compile a consensus picture. In general, the more remote the time, the more likely the story is incomplete or in error.

Life evolved in the sea. It stayed there for the majority of the history of earth. The first replicating molecules were most likely RNA. RNA is a nucleic acid similar to DNA. In laboratory studies it has been shown that some RNA sequences have catalytic capabilities. Most importantly, certain RNA sequences act as polymerases — enzymes that form strands of RNA from its monomers. This process of self-replication is the crucial step in the formation of life. This is called the RNA world hypothesis. The common ancestor of all life probably used RNA as its genetic material. This ancestor gave rise to three major lineages of life. These are: the prokaryotes ("ordinary" bacteria), archaebacteria (thermophilic, methanogenic and halophilic bacteria) and eukaryotes. Eukaryotes include protists (single celled organisms like amoebas and diatoms and a few multicellular forms such as kelp), fungi (including mushrooms and yeast), plants and animals.

Eukaryotes and archaebacteria are the two most closely related of the three. The process of translation (making protein from the instructions on a messenger RNA template) is similar in these lineages, but the organization of the genome and transcription (making messenger RNA from a DNA template) is very different in prokaryotes than in eukaryotes and archaebacteria. Scientists interpret this to mean that the common ancestor was RNA based; it gave rise to two lineages that independently formed a DNA genome and hence independently evolved mechanisms to transcribe DNA into RNA. The first cells must have been anaerobic because there was no oxygen in the atmosphere. In addition, they were probably thermophilic ("heat-loving") and fermentative. Rocks as old as 3.5 billion years old have yielded prokaryotic fossils. Specifically, some rocks from Australia called the Warrawoona series give evidence of bacterial communities organized into structures called stromatolites. Fossils like these have subsequently been found all over the world. These mats of bacteria still form today in a few locales (for example, Shark Bay Australia). Bacteria are the only life forms found in the rocks for a long, long time —eukaryotes (protists) appear about 1.5 billion years ago and fungi-like things appear about 900 million years ago (0.9 billion years ago).

Photosynthesis evolved around 3.4 billion years ago. Photosynthesis is a process that allows organisms to harness sunlight to manufacture sugar from simpler precursors. The first photosystem to evolve, PSI, uses light to convert carbon dioxide (CO_2) and hydrogen sulfide (H_2S) to glucose. This process releases sulfur as a waste product. About a billion years later, a second photosystem (PS) evolved, probably from a duplication of the first photosystem. Organisms with PSII use both photosystems in conjunction to convert carbon dioxide (CO_2) and water (H_2O) into glucose.

This process releases oxygen as a waste product. Anoxygenic (or H_2S) photosynthesis, using PSI, is seen in living purple and green bacteria. Oxygenic (or H_2O) photosynthesis, using PSI and PSII, takes place in cyanobacteria. Cyanobacteria are closely related to and hence probably evolved from purple bacterial ancestors. Green bacteria are an outgroup. Since oxygenic bacteria are a lineage within a cluster of anoxygenic lineages, scientists infer that PSI evolved first. This also corroborates with geological evidence.

Green plants and algae also use both photosystems. In these organisms, photosynthesis occurs in organelles (membrane bound structures within the cell) called chloroplasts. These organelles originated as free living bacteria related to the cyanobacteria that were engulfed by ur-eukaryotes and eventually entered into an endosymbiotic relationship. This endosymbiotic theory of eukaryotic organelles was championed by Lynn Margulis. Originally controversial, this theory is now accepted. One key line of evidence in support of this idea came when the DNA inside chloroplasts was sequenced — the gene sequences were more similar to free-living cyanobacteria sequences than to sequences from the plants the chloroplasts resided in.

After the advent of photosystem II, oxygen levels increased. Dissolved oxygen in the oceans increased as well as atmospheric oxygen. This is sometimes called the oxygen holocaust. Oxygen is a very good electron acceptor and can be very damaging to living organisms. Many bacteria are anaerobic and die almost immediately in the presence of oxygen. Other organisms, like animals, have special ways to avoid cellular damage due to this element (and in fact require it to live.) Initially, when oxygen began building up in the environment, it was neutralized by materials already present. Iron, which existed in high concentrations in the sea was oxidized and precipitated. Evidence of this can be seen in banded iron formations from this time, layers of iron deposited on the sea floor. As one geologist put it, "the world rusted." Eventually, it grew to high enough concentrations to be dangerous to living things. In response, many species went extinct, some continued (and still continue) to thrive in anaerobic microenvironments and several lineages independently evolved oxygen respiration.

The purple bacteria evolved oxygen respiration by reversing the flow of molecules through their carbon fixing pathways and modifying their electron transport chains. Purple bacteria also enabled the eukaryotic lineage to become aerobic. Eukaryotic cells have membrane bound organelles called mitochondria that take care of respiration for the cell. These are endosymbionts like chloroplasts. Mitochondria formed this symbiotic relationship very early in eukaryotic history, all but a few groups of eukaryotic cells have mitochondria. Later, a few lineages picked up chloroplasts. Chloroplasts have multiple origins. Red algae picked up ur-chloroplasts from the cyanobacterial lineage. Green algae, the group plants evolved from, picked up different urchloroplasts from a prochlorophyte, a lineage closely related to cyanobacteria.

Animals start appearing prior to the Cambrian, about 600 million years ago. The first animals dating from just before the Cambrian were found in rocks near Adelaide, Australia. They are called the Ediacarian fauna and have subsequently been found in other locales as well. It is unclear if these forms have any surviving descendants. Some look a bit like Cnidarians (jellyfish, sea anemones and the like); others resemble annelids (earthworms). All the phyla (the second highest taxonomic category) of animals appeared around the Cambrian. The Cambrian 'explosion' may have been a result of higher oxygen concentrations enabling larger organisms with higher metabolisms to evolve. Or it might be due to the spreading of shallow seas at that time providing a variety of new niches. In any case, the radiation produced a wide variety of animals.

Some paleontologists think more animal phyla were present then than now. The animals of the Burgess shale are an example of Cambrian animal fossils. These fossils, from Canada, show a bizarre array of creatures, some which appear to have unique body plans unlike those seen in any living animals. The extent of the Cambrian explosion is often overstated. Although quick, the Cambrian explosion is not instantaneous in geologic time. Also, there is evidence of animal life prior to the Cambrian. In addition, although all the phyla of animals came into being, these were not the modern forms we see today. Our own phylum (which we share with other mammals, reptiles, birds, amphibians and fish) was represented by a small, sliver-like thing calledPikaia. Plants were not yet present. Photosynthetic protists and algae were the bottom of the food chain. Following the Cambrian, the number of marine families leveled off at a little less than 200.

The Ordovician explosion, around 500 million years ago, followed. This 'explosion', larger than the Cambrian, introduced numerous families of the Paleozoic fauna (including crinoids, articulate brachiopods, cephalopods and corals). The Cambrian fauna, (trilobites, inarticulate brachiopods, etc.) declined slowly during this time. By the end of the Ordovician, the Cambrian fauna had mostly given way to the Paleozoic fauna and the number of marine families was just over 400. It stayed at this level until the end of the Permian period. Plants evolved from ancient green algae over 400 million years ago. Both groups use chlorophyll a and b as photosynthetic pigments. In addition, plants and green algae are the only groups to store starch in their chloroplasts. Plants and fungi (in symbiosis) invaded the land about 400 million years ago.

The first plants were moss-like and required moist environments to survive. Later, evolutionary developments such as a waxy cuticle allowed some plants to exploit more inland environments. Still mosses lack true vascular tissue to transport fluids and nutrients. This limits their size since these must diffuse through the plant. Vascular plants evolved from mosses. The first vascular land plant known is Cooksonia, a spiky, branching, leafless structure. At the same time, or shortly thereafter, arthropods followed plants onto the land. The first land animals known are myriapods — centipedes and millipedes. Vertebrates moved onto the land by the Devonian period, about 380 million years ago. Ichthyostega, an amphibian, is the among the first known land vertebrates.

It was found in Greenland and was derived from lobe-finned fishes called Rhipidistians. Amphibians gave rise to reptiles. Reptiles had evolved scales to decrease water loss and a shelled egg permitting young to be hatched on land. Among the earliest well preserved reptiles is Hylonomus, from rocks in Nova Scotia. The Permian extinction was the largest extinction in history. It happened about 250 million years ago. The last of the Cambrian Fauna went extinct. The Paleozoic fauna took a nose dive from about 300 families to about 50. It is estimated that 96% of all species (50% of all families) met their end. Following this event, the Modern fauna, which had been slowly expanding since the Ordovician, took over. The Modern fauna includes fish, bivalves, gastropods and crabs. These were barely affected by the Permian extinction. The Modern fauna subsequently increased to over 600 marine families at present. The Paleozoic fauna held steady at about 100 families. A second extinction event shortly following the Permian kept animal diversity low for awhile. During the Carboniferous (the period just prior to the Permian) and in the Permian the landscape was dominated by ferns and their relatives. After the Permian extinction, gymnosperms (ex. pines) became more abundant. Gymnosperms had evolved seeds, from seedless fern ancestors, which helped their ability to disperse. Gymnosperms also evolved pollen, encased sperm which allowed for more outcrossing.

Dinosaurs evolved from archosaur reptiles, their closest living relatives are crocodiles. One modification that may have been a key to their success was the evolution of an upright stance. Amphibians and reptiles have a splayed stance and walk with an undulating pattern because their limbs are modified from fins. Their gait is modified from the swimming movement of fish. Splay stanced animals cannot sustain continued locomotion because they cannot breathe while they move; their undulating movement compresses their chest cavity. Thus, they must stop every few steps and breath before continuing on their way. Dinosaurs evolved an upright stance similar to the upright stance mammals independently evolved.

This allowed for continual locomotion. In addition, dinosaurs evolved to be warm-blooded. Warm-bloodedness allows an increase in the vigour of movements in erect organisms. Splay stanced organisms would probably not benefit from warm- bloodedness. Birds evolved from sauriscian dinosaurs. Cladistically, birds are dinosaurs. The transitional fossil Archaeopteryx has a mixture of reptilian and avian features. Angiosperms evolved from gymnosperms, their closest relatives are Gnetae. Two key adaptations allowed them to displace gymnosperms as the dominant fauna — fruits and flowers. Fruits (modified plant ovaries) allow for animal-based seed dispersal and deposition with plenty of fertilizer.

Flowers evolved to facilitate animal, especially insect, based pollen dispersal. Petals are modified leaves. Angiosperms currently dominate the flora of the world — over three fourths of all living plants are angiosperms. Insects evolved from primitive segmented arthropods. The mouth parts of insects are modified legs. Insects are closely related to annelids. Insects dominate the fauna of the world. Over half of all named species are insects. One-third of this number are beetles. The end of the Cretaceous, about 65 million years ago, is marked by a minor mass extinction. This extinction marked the demise of all the lineages of dinosaurs save the birds. Up to this point mammals were confined to nocturnal, insectivorous niches. Once the dinosaurs were out of the picture, they diversified. Morgonucudon, a contemporary of dinosaurs, is an example of one of the first mammals. Mammals evolved from therapsid reptiles. The finback reptile Diametrodon is an example of a therapsid.

One of the most successful lineages of mammals is, of course, humans. Humans are neotenous apes. Neoteny is a process which leads to an organism reaching reproductive capacity in its juvenile form. The primary line of evidence for this is the similarities between young apes and adult humans. Louis Bolk compiled a list of 25 features shared between adult humans and juvenile apes, including facial morphology, high relative brain weight, absence of brow ridges and cranial crests. The earth has been in a state of flux for 4 billion years. Across this time, the abundance of different lineages varies wildly. New lineages evolve and radiate out across the face of the planet, pushing older lineages to extinction, or relictual existence in protected refugia or suitable microhabitats.

Organisms modify their environments. This can be disastrous, as in the case of the oxygen holocaust. However, environmental modification can be the impetus for further evolutionary change. Overall, diversity has increased since the beginning of life. This increase is, however, interrupted numerous times by mass extinctions. Diversity appears to have hit an all-time high just prior to the appearance of humans. As the human population has increased, biological diversity has decreased at an ever-increasing pace. The correlation is probably causal.

Scientific Standing of Evolution and its Critics

The theory of evolution and common descent were once controversial in scientific circles. This is no longer the case. Debates continue about how various aspects of evolution work. For example, all the details of patterns of relationships are not fully worked out. However, evolution and common descent are considered fact by the scientific community.

Scientific creationism is 100% crap. So-called "scientific" creationists do not base their objections on scientific reasoning or data. Their ideas are based on religious dogma, and their approach is simply to attack evolution. The types of arguments they use fall into several categories: distortions of scientific principles, straw man versions of evolution, dishonest selective use of data appeals to emotion or wishful thinking, appeals to personal incredulity, dishonestly quoting scientists out of context and simply fabricating data to suit their arguments. Most importantly, scientific creationists do not have a testable, scientific theory to replace evolution with. Even if evolution turned out to be wrong, it would simply be replaced by another scientific theory. Creationists do not conduct scientific experiments, nor do they seek publication in peer-reviewed scientific journals. Much of their output is "preaching to the choir."

The most persuasive creationist argument is a non-scientific one — the appeal to fair play. "Shouldn't we present both sides of the argument?," they ask. The answer is no — the fair thing to do is exclude scientific creationism from public school science courses. Scientists have studied and tested evolution for 150 years. There is voluminous evidence for it. Within the scientific community, there are no competing theories. Until scientific creationists formulate a scientific theory, and submit it for testing, they have no right to demand equal time in science class to present their ideas. Evolution has earned a place in the science curriculum. Creationism has not.

Science is based on an open and honest look at the data. Much of creationism is built on dishonest debating techniques and special pleading for a case the data does not support. Science belongs in science classes. Evolution is science. Creationism is not. It's that simple. The creationist attack on public school education means that school children are denied the possibility of learning about the most powerful and elegant theory in biology. Politicians are willing to allow the scientifically ignorant, but politically strong, to wreck the educational system in exchange for votes. People interested in evolution, and science education in general, need to closely watch school board elections. Creationist "stealth" candidates have been elected in several regions. Thankfully, many have been voted out once their views became apparent.

The majority of Americans are religious, but only a minority are religious nuts. The version of religion the far right wants to impose on America is as repulsive to most mainstream Christians as it is to members of other religions, atheists and agnostics. Most informed religious people see no reason for biological facts and theories to interfere with their religious beliefs.

PERSPECTIVES OF DEVELOPMENT BIOLOGY

Developmental biology is the study of the process by which organisms grow and develop. Modern developmental biology studies the genetic control of cell growth, differentiation and "morphogenesis," which is the process that gives rise to tissues, organs and anatomy. Developmental biology is that branch of life science, which deals with the study of the process by which organisms grow and develop. Embryology is a subfield, the study of organisms between the one-cell stage (generally, the zygote) and the end of the embryonic stage. Embryology was originally a more descriptive science until the 20th century. Embryology and developmental biology today deal with the various steps necessary for the correct and complete formation of the body of a living organism. The related field of evolutionary developmental biology was formed largely in the 1990s and is a synthesis of findings from molecular developmental biology and evolutionary biology which considers the diversity of organismal form in an evolutionary context.

Animal development is a spectacular process and represents a masterpiece of temporal and spatial control of gene expression. Developmental genetics is a very helpful process. It studies the effect that genes have in a phenotype. The findings of developmental biology can help to understand developmental malfunctions such as chromosomal aberrations, for example, Down syndrome. An understanding of the specialization of cells during embryogenesis may yield information on how to specialize stem cells to specific tissues and organs, which could lead to the specific cloning of organs for medical purposes.

Another biologically important process that occurs during development is apoptosis—programmed cell death or "suicide". For this reason, many developmental models are used to elucidate the physiology and molecular basis of this cellular process. Similarly, a deeper understanding of developmental biology can foster greater progress in the treatment of congenital disorders and diseases, *e.g.* studying human sex determination can lead to treatment for disorders such ascongenital adrenal hyperplasia.

Developmental model organisms

Often used model organisms in developmental biology include the following:

- Vertebrates
  - Zebrafish Danio rerio
  - Medakafish Oryzias latipes
  - Fugu (pufferfish) Takifugu rubripes
  - Frog Xenopus laevis, Xenopus tropicalis
  - Chicken Gallus gallus
  - Mouse Mus musculus (Mammalian embryogenesis)
- Invertebrates
  - Lancelet Branchiostoma lanceolatum
  - Ascidian Ciona intestinalis
  - Sea urchin Strongylocentrotus purpuratus
  - Roundworm Caenorhabditis elegans
  - Fruit fly Drosophila melanogaster (Drosophila embryogenesis)
- Plants (Plant embryogenesis)
  - Arabidopsis thaliana
  - Maize

  - Snapdragon Antirrhinum majus
- Other
  - Slime mold Dictyostelium discoideum

Studied phenomena

Cell Differentiation

Differentiation is the formation of cell types, from what is originally one cell – the zygote or spore. The formation of cell types like nerve cells occurs with a number of intermediary, less differentiated cell types. A cell stays a certain cell type by maintaining a particular pattern of gene expression. This depends on regulatory genes, *e.g.* for transcription factors and signaling proteins.

These can take part in self-perpetuating circuits in the gene regulatory network, circuits that can involve several cells that communicate with each other. External signals can alter gene expression by activating a receptor, which triggers a signaling cascade that affects transcription factors. For example, the withdrawal of growth factors from myoblastscauses them to stop dividing and instead differentiate into muscle cells.

Embryonal Development

Embryogenesis is the step in the life cycle after fertilisation – the development of the embryo, starting from the zygote(fertilised egg). Organisms can differ drastically in the how embryo develops, especially when belong to different phyla. For example, embryonal development in placental mammals starts with cleavage of the zygote into eight uncommited cells, which then form a ball (morula). The outer cells become the trophectoderm which will form the fetal part of the placenta, while inner cells become the inner cell mass that will form all other organs.

In contrast, the fruit fly zygote first forms a sausage-shaped syncytium, which is still one cell but with many cell nuclei. Patterning is important for determining which cells develop which organs. This is mediated by signaling between adjacent cells by proteins on their surfaces, and by gradients of signaling molecules. An example is retinoic acid, which forms a gradient in the head to tail direction in animals.

Retinoic acid enters cells and activates Hox genes in a concentration-dependent manner – Hox genes differ in how much retinoic acid they require for activation. As Hox genes code for transcription factors, this causes discrete segments in the head to tail direction. This is important for the segmentation of the spine in vertebrates. Embryonal development does not always go right, and errors can result in birth defects or miscarriage. Often the reason is genetic (mutation or chromosome abnormality), but there can be environmental influence (teratogens).Abnormal development is also of evolutionary interest as it provides a mechanism for changes in body play).

Growth

Growth is the enlargement of a tissue or organism. Growth continues after the embryonal stage, and occurs through cell proliferation, enlargement of cells or accumulation of extracellular material. In plants, growth results in an adult organism that is strikingly different from the embryo. The proliferating cells tend to be distinct from differentiated cells.

In some tissues proliferating cells are restricted to specialised areas, such as the growth plates of bones. But some stem cells migrate to where they are needed, such as mesenchymal stem cells which can migate from the bone marrow to form *e.g.* muscle, bone or adipose tissue.

The size of an organ frequently determines its growth, as in the case of the liver which grows back to its previous size if a part is removed. Growth factors, such as fibroblast growth factors in the animal embryo and growth hormone in juvenile mammals, also control the extent of growth.

Metamorphosis

Most animals have a larval stage, with a body plan different from that of the adult organism. The larva abrubtly develops into an adult in a process called metamorphosis.

For example, caterpillars (butterfly larvae) are specilised for feeding whereas adult butterflies (imagos) are specilised for flight and reproduction. When the caterpillar has grown enough, it turns into an immobile pupa. Here, the imago develops from imaginal discs found inside the larva.

Regeneration

Regeneration is the reactivation of development so that a missing body part grows back. This phenomenon has been studied particularly in salamanders, where the adults can reconstruct a whole limb after it has been amputated. Researchers hope to one day be able to induce regeneration in humans·

There is little spontaneous regeneration in adult humans, although the liver is a notable exception. Like for salamanders, the regeneration of the liver involves dedifferentiation of some cells to a more embryonal state.

THE IMPORTANCE OF EVOLUTION IN BIOLOGY

"Nothing in biology makes sense except in the light of evolution." – Theodosius Dobzhansky. Evolution has been called the cornerstone of biology, and for good reasons. It is possible to do research in biology with little or no knowledge of evolution. Most biologists do. But, without evolution biology becomes a disparate set of fields. Evolutionary explanations pervade all fields in biology and brings them together under one theoretical umbrella. We know from microevolutionary theory that natural selection should optimize the existing genetic variation in a population to maximize reproductive success.

This provides a framework for interpreting a variety of biological traits and their relative importance. For example, a signal intended to attract a mate could be intercepted by predators. Natural selection has caused a trade-off between attracting mates and getting preyed upon. If you assume something other than reproductive success is optimized, many things in biology would make little sense. Without the theory of evolution, life history strategies would be poorly understood. Macroevolutionary theory also helps explain many things about how living things work. Organisms are modified over time by cumulative natural selection.

The numerous examples of jury-rigged design in nature are a direct result of this. The distribution of genetically based traits across groups is explained by splitting of lineages and the continued production of new traits by mutation. The traits are restricted to the lineages they arise in. Details of the past also hold explanatory power in biology.

Plants obtain their carbon by joining carbon dioxide gas to an organic molecule within their cells. This is called carbon fixation. The enzyme that fixes carbon is RuBP carboxlyase. Plants using C3 photosynthesis lose 1/3 to 1/2 of the carbon dioxide they originally fix. RuBP carboxlyase works well in the absence of oxygen, but poorly in its presence.

This is because photosynthesis evolved when there was little gaseous oxygen present. Later, when oxygen became more abundant, the efficiency of photosynthesis decreased. Photosynthetic organisms compensated by making more of the enzyme. RuBP carboxylase is the most abundant protein on the planet partially because it is one of the least efficient. Ecosystems, species, organisms and their genes all have long histories.

A complete explanation of any biological trait must have two components. First, a proximal explanation – how does it work? And second, an ultimate explanation – what was it modified from? For centuries humans have asked, "Why are we here?" The answer to that question lies outside the realm of science. Biologists, however, can provide an elegant answer to the question, "How did we get here?"

LIFE

Life is an attribute of objects that are considered alive. There are a vast variety of living organisms. Properties common to these organisms – plants, animals, fungi, protists, archaea and bacteria – are being carbon; water-based, and being cellular with complex organization. They undergo metabolism; possess a capacity to grow; respond to stimuli, reproduce and through natural selection adapt to their environment in succeeding generations. An entity with the above properties is considered to be an organism. However, not every definition of life considers all of these properties to be essential. For example, the capacity for descent with modification is often taken as the only essential property of life. This definition notably includes viruses, which do not qualify under narrower definitions as they are acellular and do not metabolise. Broader definitions of life may also include theoretical non-carbon-based life and other alternative biology.

There is no universal definition of life; there are a variety of definitions proposed by different scientists.

A conventional definition: Often scientists say that life is a characterstic of organisms that exhibit the following phenomena:

Homeostasis: Regulation of the internal environment to maintain a constant state; for example, sweating to reduce temperature.

Organization: Being composed of one or more cells, which are the basic units of life.

Metabolism: Consumption of energy by converting nonliving material into cellular components (anabolism) and decomposing organic matter (catabolism). Living things require energy to maintain internal organization (homeostasis) and to produce the other phenomena associated with life.

Growth: Maintenance of a higher rate of synthesis than catalysis. A growing organism increases in size in all of its parts, rather than simply accumulating matter. The particular species begins to multiply and expand as the evolution continues to flourish.

Adaptation: The ability to change over a period of time in response to the environment. This ability is fundamental to the process of evolution and is determined by the organism's heredity as well as the composition of metabolized substances, and external factors present.

Response to stimuli: A response can take many forms, from the contraction of a unicellular organism when touched to complex reactions involving all the senses of higher animals. A response is often expressed by motion, for example, the leaves of a plant turning toward the sun or an animal chasing its prey.

Reproduction: The ability to produce new organisms. Reproduction can be the division of one cell to form two new cells. Usually the term is applied to the production of a new individual (either asexually, from a single parent organism, or sexually, from at least two differing parent organisms), although strictly speaking it also describes the production of new cells in the process of growth.

However, others cite several limitations of this definition. Thus, many members of several species do not reproduce, possibly because they belong to specialized sterile castes (such as ant workers), these are still considered forms of life. One could say that the property of life is inherited; hence, sterile hybrid species are considered life although not themselves capable of reproduction. It is also worth noting that non-reproducing individuals may still help the spread of their genes through such mechanisms as kin selection.

Viruses and aberrant prion proteins are often considered replicators rather than forms of life, a distinction warranted because they cannot reproduce without very specialized substrates such as host cells or proteins, respectively. However, most forms of life rely on foods produced by other species, or at least the specific chemistry of Earth's environment.

Still others contest such definitions of life on philosophical grounds. They offer the following as examples of life: viruses which reproduce; storms or flames which "burn"; certain computer software programmes which are programmed to mutate and evolve; future software programmes which may evince (even high-order) behaviour; machines which can move; and some forms of proto-life consisting of metabolizing cells without the ability to reproduce. Still, most scientists would not call such phenomena expressive of life. Generally all seven characteristics are required for a population to be considered a life form.

The systemic definition of life is that living things are self-organizing and autopoietic (self-producing). These objects are not to be confused with dissipative structures (e.g. fire).

Variations of this definition include Stuart Kauffman's definition of life as an autonomous agent or a multi-agent system capable of reproducing itself or themselves, and of completing at least one thermodynamic work cycle.

Yet other definitions of life are:

Living things are systems that tend to respond to changes in their environment, and inside themselves, in such a way as to promote their own continuation.

Life is a charateristic of self-organizing, cannibalistic systems consisting of a population of replicators that are capable of mutation, around most of which homeostatic, metabolizing organisms evolve. This definition does not include flames, but does include worker ants, viruses and mules. Self reproduction and energy consumption is only one means for a system to promote its own continuation. This explains why bees can be alive and yet commit suicide in defending their hive. In this case the whole colony works as such a living system.

Type of organization of matter producing various interacting forms of variable complexity, whose main property is to replicate almost perfectly by using matter and energy available in their environment to which they may adapt. In this definition "almost perfectly" relates to mutations happening during replication of organisms that may have adaptative benefits.

Contradictions in Definitions

Most of the definitions are focused on differentiating between living material and non-living material. It is still not clear whether there exists computer software that can be considered as a living. Some computer software show certain resemblance with life. It is not difficult in algorithmic sense to create software which fulfills some base criteria (growing, reproduction, reaction to the changes in the environment, etc.).

Consider an artificial environment with certain properties (for example an system which is calling the processes with different input). If the process is able to solve the problem (respond with a correct output) in a given time, it can survive, otherwise it is deleted. If the process solves the problem before the other processes it gets higher part from the computer resources. This is a very simple model of an environment in which the life has been created. More sophisticated environments can be created. Although in most of the simulations the systems collapse relatively fast (for example non of the processes are able to adapt and solve the given problem) it has not yet been formally proved that such an environment can not exist permanently. Such an environment could met all the criteria of some simple life definitions.

If ever such a theorem will be formally proved, it will be necessary to reconsider the definitions or admit that the in some sense the computer world is a good approximation of the real one.

ORIGIN OF LIFE

Although it cannot be pinpointed exactly, evidence suggests that life on Earth has existed for about 3.7 billion years .There is no truly "standard" model for the origin of life, but most currently accepted scientific models build in one way or another on the following discoveries, which are listed roughly in order of postulated emergence:

Plausible pre-biotic conditions result in the creation of the basic small molecules of life. This was demonstrated in the Miller-Urey experiment.

Phospholipids spontaneously form lipid bilayers, the basic structure of a cell membrane.

Procedures for producing random RNA molecules can produce ribozymes, which are able to produce more of themselves under very specific conditions.

There are many different hypotheses regarding the path that might have been taken from simple organic molecules to protocells and metabolism. Many models fall into the "genes-first" category or the "metabolism-first" category, but a recent trend is the emergence of hybrid models that do not fit into either of these categories.

EXTRATERRESTRIAL LIFE

Earth is the only planet in the universe known to harbour life. The Drake equation has been used to estimate the probability of life elsewhere, but scientists disagree on many of the values of variables in this equation (although strictly speaking Drake equation estimates the number of extraterrestrial civilizations in our galaxy with which we might come in contact - not probability of life elsewhere). Depending on those values, the equation may either suggest that life arises frequently or infrequently.

ORIGIN OF LIFE CONCEPTS

In the physical sciences, abiogenesis, the question of the origin of life, is the study of how life on Earth might have evolved from non-life sometime between 3.9 and 3.5 billion years ago. This topic also includes theories and ideas regarding possible extra-planetary or extra-terrestrial origin of life hypotheses, thought to have possibly occurred over the last 13.7 billion years in the evolution of the known universe since the big bang.

Origin of life studies is a limited field of research despite its profound impact on biology and human understanding of the natural world. Progress in this field is generally slow and sporadic, though it still draws the attention of many due to the eminence of the question being investigated. A few facts give insight into the conditions in which life may have emerged, but the mechanisms by which non-life became life are still elusive.

Asking the question about different theories on the origin of life is similar to asking the question about the possible ways other entities like a house, a car, or a factory could have come into existence. There are only three possible ways that such things could have come into existence. They are as follows:

- Random, natural unguided forces or processes
- Designed and created by a designer
- Self generation or ability to create, inherent or designed into matter.

Arguments that life spores were transported to our planet, the earth, still requires answering the same question. Which possible theory was the correct mechanism to explain the generation of life at the remote location where the spores originated?

We all know that a car or a factory is designed and created by a designer. Nobody would question that. What we have learned is that a car is both far less complex than a single living cell and cannot reproduce itself like some living cells can. Yet some of us are willing to consider that such living cells came about by a random unguided process from inorganic matter.

A very modern complex automated factory is more similar to the complexity and function of a living cell. Although it is considerably less complex, much larger and unable to reproduce itself, at least it can produce whatever it was designed to produce. Yet nobody would believe that such a factory could come into existence by a random unguided process. Why do we believe that life could have come into existence by a random unguided process?

Even if we have the absurd belief that an extremely complex living cell could have came into existence through random processes, how much more absurd is it to believe that a cell would come into existence that could reproduce itself. None of us believe that a complex factory could come into existence by a random process. How much more absurd is it that a factory that could reproduce itself could have come into existence by random chance? Man with his great intelligence has not considered designing a factory that could reproduce itself even on a macroscopic scale. How much more intelligence is required to create life with extremely complex, "machine like" cells that can reproduce themselves on a microscopic level.

Some evolutionists speculate that matter has inherent or creative power built into it. If random chance cannot produce non-reproductive life from matter (nobody has shown that it can) how can the much, much more complex inherent creative ability of matter be credited to random chance?

Astrophysicist, Sir Fred Hoyle and his colleague Chandra Wickramasinghe argued chance processes could not have formed the biochemical machinery of the cell, especially the enzymes. In their book, "Evolution from Space," they estimated the probability of forming a single enzyme of protein at random, in the rich ocean of amino acids, was no more than 10 to the 20th power. They then calculated the likelihood of forming by chance all of the more than 2000 enzymes used in the life forms on earth. This probability was calculated at one in 10 to the 40,000th power.

A vivid analogy from Hoyle became a well-known cliché. "Belief in chemical evolution of the first cell from lifeless chemicals is equivalent to believing that a tornado could sweep through a junkyard and form a Boeing 747."

Even with all the correct chemicals present, life cannot be created. If we chopped some existing form of life in a blender, life cannot be generated from all those correct chemicals. Even if we have a real living configuration, we cannot recreate life. Dead or frozen bodies cannot be brought back to life after an extended lifeless time.

We need to accept the explanation that God created the universe and all life and direct our beliefs and actions to be consistent with the acknowledgment of the creator and Jesus' revelation that He is the way, the truth, and the life.

Recent rapid advances in the study of genetics and molecular biology have produced additional insight into fundamental questions such as the origin of life on Earth. This paper provides an overview of Genetics from an engineering perspective as well as discussions of these questions.

The striking similarities are present between genetics processes and computer science. There is a digital genetic code. There is digital logic with "and" functions and logic matrices. There is even "error correcting" in digital copying of genetic code.

GENETICS AND ORIGIN OF LIFE

All living organisms have a genetic code generally represented by the sequence of nucleotides in their DNA. Since there are four possible bases used in construction of the code (denoted A, G, C, and T), each letter in the code carries two bits of information. The DNA is normally in the form of a double strand (the famous "double helix") where the second strand is complementary to the first strand. That is, in the second strand a sequence such as "AGCTTT" is replaced by "TCGAAA" which carries the same information. So the terminology "base pair" refers to one letter of genetic code represented by the base and its complement and equivalent to two bits of information in computer parlance.

Humans have a genetic code (the "genome") of about 3.3 billion base pairs (6.6 gigabits or 825 megabytes). Yes, the human genome would easily fit on a typical laptop hard drive. Each human has two copies of the genome in virtually every cell of his or her body because of inheritance of one set each from both father and mother. Actually males have about 2 percent less code in one of their genomes because they only have one "X" chromosome. During growth and normal life processes cells endlessly read, and interpret the genetic codes, copy various snippets of the code, and use the copies as templates in the manufacture of proteins.

Early naturalists thought that genetic traits were inherited in more or less "analog" fashion in which offspring had an average of their parent's characteristics. Gregor Mendel was the first to realize through extensive experiments with breeding of peas that at the lowest level, inheritance is binary, and that there is a minimum unit of inheritance now known as a "gene". Mendel found that some traits are "recessive" and can appear in progeny even if neither parent has the trait. He also found that inheritance of a trait is independent of inheritance of other traits.

Genes are now known to be implemented as sequences of genetic code that direct specific cells to produce a particular protein at a particular time. An essentially infinite number of possible different protein molecules can be produced depending on the particular order of amino acid molecules used in their construction. The code for protein production has been "broken" so that we now know that a three-letter sequence (a codon) is used to specify a particular amino acid (there are 20 amino acids). For instance, the sequence GGC specifies that the amino acid glycine is to be added to a protein molecule. Start and stop codons mark the beginning and end of a protein coding sequence in a manner startlingly like modern data communications schemes. There are 64 possible codons and only 20 possible amino acids so some redundancy and error correction exists. The regulatory code sequences in genes that specify in which parts of the body and/or at which times a protein will be produced are much more complex and less well understood.

For humans, approximately 45,000 genes are contained in 23 separate strands of DNA known as chromosomes (46 if both sets of code are counted). The number of chromosomes is not indicative of complexity. Dogs have 78; Horses have 64. Ferns have 512. The international Human Genome Project

(HGP) has completed a preliminary "draft" sequencing of the entire human genome genetic code. Sequences of a small number of other organisms such as the mouse, fruit fly, and e coli are completed or in work. Having the sequence is very different from understanding what it means.

Mendelian Genetics is somewhat like Newtonian Physics. Eventually scientists noticed subtle deviations from the inheritance model predicted by Mendel. Specifically, inheritance of certain traits was not completely independent of other traits. We now know that inheritance of traits will be independent only if they are carried by different chromosomes and that the probability of jointly inheriting traits carried by the same chromosome is proportional to the physical distance between the two genes on the chromosome. Extensive inheritance studies have resulted in maps of the genome showing the approximate location of some trait genes and human genetic disease genes on specific chromosomes. This information can eventually be combined with the detailed sequence data to disclose the genes which (when incorrect) are responsible for genetic diseases. There are an estimated 3000 different human genetic diseases.

Gene Logic

The genetic code has been compared to a blueprint specifying the design of an organism. In fact the genetic code specifies not only the design of the organism but provides for the mechanisms needed to "read" the code and manufacture the components of the organism as well as specifying the procedures needed for the life processes of the finished organism. Simple organisms are completely defined genetically. Each tiny nematode worm has exactly 958 cells. Humans, on the other hand, have trillions of cells and less than 100,000 genes so the genetic code is more of a general plan. For example, major blood vessels are genetically specified. Everybody has an aorta. But minor blood vessels grow where needed according to genetically defined rules.

Although all the somatic cells in an organism contain the complete genetic code, in any given cell only a relatively few genes are active. The difference in the genes that are active determines the difference between, say, liver and brain cells. A complex gene logic determines when and where a particular gene will be "turned on". The gene logic can accommodate varying amounts of positional detail. The eye, which has a complex structure in which adjacent cells can be very different, presumably requires many genes to implement a relatively small structure. The femur is much larger but much less complex and requires less genetic information. The gene logic also controls when various activities will take place. Cells divide rapidly in growing organisms but do not divide in adults unless needed to replace dead or discarded cells. (Cancer involves a major breakdown in the gene logic in which cells grow in both an inappropriate position and at an inappropriate time. Cancer is thought to require multiple mutations, some of which can be inherited.)

The gene logic is implemented using signaling proteins. That is, genes can control the production of proteins which are actually the building blocks to produce muscle and other structural components in growing cells but can also control production of other proteins which are logic signals. These logic signals can then be received by other genes and determine whether those genes are activated. Some logic proteins are long range in that they can travel through essentially the entire organism (think insulin). Other shorter range signals appear only near their point of origin, possibly only immediately around the cell in which they are generated. Since genes can both generate and detect signaling proteins, many genes can implement a very complex logic. The positional logic framework which governs where in the body specific types of cells are found is an example of a "boot-strap" problem. The logic framework itself has to be constructed as the organism grows from a one-cell fertilized egg to an adult.

A mutation occurs when the genetic code in a cell is altered such that descendent cells formed by division of the altered cell also have the altered DNA. If the mutation occurs in the chain of cell division between the original fertilized egg and reproductive (sperm or egg) cells (the germ line), then the mutation can be passed to progeny. Many other mutations presumably have no effect because they occur in genes that are never activated in the descendents of the affected cells. (A mutation in a gene that was only active in the brain would have no effect if it occurred in the line of cells that were to form a leg, etc)

Evolution takes place by means of mutations which affect the germ line. Often a mutation results in loss of some essential function and is therefore fatal to progeny and not passed on to living descendants. Sometimes the mutation results in an evolutionary advantage and therefore may eventually become universal in descendants. Sometimes mutation results in characteristics which are different (such as a red eye colour in a species that previously had only brown eyes) but confers no particular advantage or disadvantage and so becomes common but not universal in descendants. Higher organisms also have extensive non-functional sections in their genetic code. Mutations in the non-functional parts of the code would have no observable effect on the organism and therefore would be passed to descendents. Since the rate at which mutations occur should be relatively constant, differences in non-functional code can be used to determine the time since two individuals shared a common ancestor.

The HGP indicates that the human genome contains about 50 percent apparently non-functional code much of which consists of many repetitions of simple sequences such as ..ATATATATATAT...which clearly have little or no information content. Some of the repeat sequences are known to be necessary synchronization patterns such as the sequences at the beginnings and ends of chromosomes. The purpose, if any, of other repeat sequences is unknown.

3

Animal Physiology

INTRODUCTION

The field of animal physiology extends the tools and methods of human physiology to non-human animal species. Plant physiology also borrows techniques from both fields. Its scope of subjects is at least as diverse as the tree of life itself. Due to this diversity of subjects, research in animal physiology tends to concentrate on understanding how physiological traits changed throughout the evolutionary history of animals.

Other major branches of scientific study that have grown out of physiology research include biochemistry, biophysics, paleobiology, biomechanics, and pharmacology. Although the various branches of human knowledge cannot be sharply delimited from each other, it has been found expedient for practical purposes to distinguish one field of study from another. Thus chemistry, although close to physics, is set apart from it, and within the broad field of chemistry one can recognize such special subjects as inorganic chemistry, qualitative analysis, physical chemistry, etc.

Not infrequently, as knowledge expands, the older subjects become subdivided. Thus electrochemistry and colloid chemistry are offshoots from physical chemistry.

From a broad philosophical point of view some of the distinctions between sciences and their various subdivisions are none too valid. New subjects appear and receive names in treatises or college catalogs, often because some specialist desires to pass on to students information which he knows best or in which he is most interested.

In spite of the fact that distinctions between various sciences and their subdivisions may be more or less arbitrary, each branch of knowledge has to a greater or less extent a reason for existence.

Perhaps one can think of each scientific subject as representing an attempt to answer a question. Thus, in biological science, anatomy covers the question of form. It tells us what organisms or their parts look like.

Physiology, on the other hand, is concerned with the problem of function. It attempts to describe the actions of various organs or tissues and, more than this, it attempts to explain this action in terms of known chemical and physical forces. In such explanation, moreover, there is an effort to relate the phenomena in terms of cause and effect. Obviously there are various branches of physiological science.

Because of its intimate relation to the science and art of medicine, human physiology has been studied since the very beginning of human learning. Our knowledge of the physiology of other forms of life has been focused on certain types rather than others. Insect physiology is now rather widely studied; plant physiology is an old and well-developed subject. Some groups of organisms have been neglected, and there is comparatively little information about the physiology of certain invertebrate phyla. Sometimes it is convenient to lump together the physiological knowledge of all sorts of organisms into a subject known as "comparative physiology."

The point of view of this subject is to emphasize the differences rather than the resemblances between various types of living material. Opinions may differ as to the proper definition of general physiology. To some it may be little more than a diluted vertebrate or human physiology. I should prefer to consider general physiology in the broadest possible sense: it is the science which deals with all forms of living material. It attempts to answer the question, "What is life?" The general physiologist believes that all life has something in common; his search is often for the least common denominator of living processes. That such a search may not be entirely futile is indicated by the recent history of biological thought.

Modern progress in biological science has resulted largely from the fortunate realization that living things, no matter how diverse they appear in externals, are fundamentally much alike. There is an underlying unity which has proved of tremendous importance. Cytologists learned to know cell division and chromosomes from a study of cells best suited for their purposes. Fertilization was studied first in the egg cells of marine invertebrates, and only later was the same process shown to occur in widely different types of organisms.

The fundamental facts of genetics were discovered first for plants, and modern study in this field for the most part contents itself with work on these forms which are easy to raise and which breed rapidly. And yet no one doubts that genetic science is capable of universal application. In physiology, on the other hand, the tendency has been to stress the study of vertebrate or mammalian tissues. Physiology is so close to medicine and its facts are so certain of practical application that there is a strong desire to study animals close to man. The field of mammalian or human physiology has had a marvelous development, and it would be absurd to belittle its striking achievements.

And yet, in spite of the fact that students of the physiology of mammals have accumulated a vast store of precise and valuable information, the ultimate mechanisms underlying the activity of mammalian cells and tissues are but little understood. The physiologist has been loath to accept the method of attack so fruitful in other branches of biology.

If students of genetics or cytology had insisted on studying mammals or vertebrates and no other organisms, these sciences would scarcely have developed far. Great advances in theoretical biology are made by study of that type of living material particularly adapted to the problem at hand. Cells of mammals or vertebrates are often more difficult to study than those of simpler forms.

The general physiologist, in his search for fundamental information, uses whatever material he can find best suited for his purpose. It is the aim of the general physiologist to discover, in so far as possible, the nature and mechanism of living matter. This is the *raison d'être* of the science. Its students are more or less convinced that, if one could discover the essential principles which govern the activities and life of any one living cell or living material, the same principles could be used, with minor adjustments and modifications, to explain the life of practically every type of cell.

This may seem almost too much to ask, but the progress of biology has repeatedly shown the deep underlying similarity of what seem at first the most diverse manifestations of the living process.

The extreme difficulty of biological research is thus to some extent compensated by the underlying unity of living materials. Although there may be a million different species of animals, there are not a million different biologies, but rather one fundamental science with variations. In general the cell is the unit of living matter, and the life of an organism is to all intents and purposes the sum total of the life of its constituent cells.

It seems clear therefore that vital machinery must be interpreted in terms of cells, and general physiology thus becomes cellular physiology. This point of view was strongly emphasized by Verworn, who states over and over again that, if physiology is to attempt fundamental explanation of vital activity, it must be a cell physiology.

There are, however, possible exceptions to this point of view. It may be argued that not all living matter is truly cellular. Thus only by a stretch of the imagination can some of the smaller bacteria be regarded as cells; and filtrable viruses which are beyond the limits of microscopic vision, and which may or may not be alive, are certainly not cells in any ordinary sense.

This argument seems to have very little force. Biologists generally realise that the cell doctrine is not absolutely true, and that there are forms of living matter which are not truly cellular. Nevertheless, the cell doctrine is true for all practical purposes, and it forms the basis for such sciences as histology, pathology, embryology, etc. Because ultramicroscopic forms of life may exist, neither the histologist nor the embryologist has seen fit to throw over the cell theory.

In the same way physicists and physical chemists cling to the second law of thermodynamics, although it is well known that there are special conditions under which the law does not hold. There is also another reason sometimes given for not identifying general physiology with cell physiology. In investigating the activity of a mass of living matter, it is often much simpler to study great numbers of cells rather than individual units. This is especially true of chemical study, for chemical properties of cells are usually additive. Thus if one is attempting to discover the change in the glycogen content of cells during activity, one can do this best by analyzing great numbers of cells.

The analysis is, therefore, conducted on tissues rather than on cells, and in this sense, from the standpoint of method or technic, the study is no longer truly cellular. But the primary interest of the investigation is the glycogen change in the individual cell, so that this argument seems rather a superficial one. At the present time, in many branches of physiology, there is an increasing tendency to study the individual cells in so far as this is possible. Muscle physiologists are giving more and more attention to the isolated single muscle fiber, and students of nerve are attempting as best they can to isolate single nerve fibers.

As a matter of fact, these recent studies of muscle and nerve physiologists have been inspired more by a desire to avoid the complexities which arise from the existence of many fibers not always in perfect time accord, rather than from any desire to study the single fiber from the viewpoint of the student of cell protoplasm. Nevertheless, they indicate a growing realization that the ultimate mechanism responsible for any form of vital activity lies inherent in the individual cells. Whether or not we regard general physiology as synonymous with cell physiology is of no great consequence. A science is whatever its devotees choose to make it. No one could be excluded from the field of general physiology because he did not think in terms of cells.

There is, however, a growing realization that an understanding of cells and cell protoplasm is essential for the interpretation of the various vital processes as they occur in animal and plant organisms. Much of the hope for the future of general physiology lies in the possibility of using simpler types of protoplasm in order to discover the general characteristics of living substance. The field of the general physiologist is as broad as all life itself. How broad this is, at the present time it is impossible to state, for there is no agreement as to the lower limits of life.

Numerous submicroscopic filtrable viruses exist, many of them causing diseases in animals and plants. But whether these viruses are alive or not is a question still being argued. True, they reproduce and some of the larger ones resemble bacteria. On the other hand, unlike bacteria they are incapable of growth on simple synthetic media and they can only exist within bacterial, plant or animal cells. Moreover, Stanley found that the virus which causes tobacco mosaic disease is a protein molecule, and since then all the plant viruses that have been studied have been found to be protein molecules.

However, although some animal viruses seem to consist wholly of protein, others like vaccinia virus contain fat and carbohydrate as well as protein. Viruses vary widely in size and shape. Some are so small that their molecular diameter is only about 25 mμ (0.000,025 mm.). The largest virus has a volume approximately 10,000 times that of the smallest. Such large viruses can be as large or larger than small bacteria. Some viruses are spherical, some are rod-shaped; one has been found to be tadpole-shaped, with a spherical head and a slender elongate tail.

In a study of these diverse, extremely tiny, living or partly living organisms or substances, much can doubtless be learned concerning the nature of life processes.

A considerable body of information has already been accumulated. General physiology is a theoretical science just as physical chemistry is. But whereas practical chemists and chemical engineers have long since learned to depend on physicochemical principles, medical scientists and students of applied human physiology and pharmacology have sometimes been slow to realise the relation of their subjects to general physiology. It is interesting to note that a leading student of pharmacology, which is one of the most practical of biological sciences, has stated that "The manner in which drugs exert their action on cells has become one of the most important fundamental problems of physiology. Fortunately, at the present time, especially among the younger medical men, there is an ever-growing realization of the importance of basic general physiology. The primary object of the general physiologist is to interpret and explain vital phenomena. Obviously the explanations he seeks must be in terms of known scientific concepts.

As best he can, the general physiologist attempts to give a complete description of living substance in terms of its physical and chemical properties; and more than that, he attempts to show exactly what physical and chemical changes occur during vital activity. But mere description is not enough. He must aim to find causal relations between the physical and chemical phenomena observable in the living material and the life itself; in other words he is interested in discovering mechanism.

This is his most difficult task. To measure mechanical, thermal, electrical or chemical change in plant and animal tissues is not too hard, but to find out exactly which changes are primarily responsible for definite phases of vital activity is always exceedingly difficult and often, at our present stage of progress, impossible. Even in the simplest types of vital activity so many changes occur that it is hard to know which ones are intimately responsible for what is happening. Similar difficulties exist even for non-living machinery.

We know how an automobile works because it has been constructed by human hands, but if we were obliged to explain the action of an automobile by what we could see, we might be inclined to believe that its movement was due to the noise it made, the heat it produced or the breeze it generated. As scientists, general physiologists are concerned only with explanations in terms of what is known or what can be known. Whether or not life can be explained in terms of known or knowable chemical and physical forces is a philosophical rather than a scientific question and it should perhaps be left to the philosophers.

For them, and for thoughtful people generally, it is a question of tremendous interest. "Since a man must needs live before he can be a philosopher, no problem of philosophy is more fundamental than the nature of life." For many years, those who believe that eventually explanations of the nature of life processes can be obtained in terms of chemistry and physics have been called mechanists. On the other hand, those who deny that physical and chemical forces can ever provide a complete explanation have been called vitalists.

Between mechanists and vitalists, the arguments have often been bitter—among working biologists generally, the term vitalist is often used as a term of reproach—and hence in recent years there has been a semantic attempt to alter the terminology or to abandon it altogether.

But the terms are still properly understood in their original meaning, and it is doubtful if other words or other definitions of the old words will make the arguments any less bitter. To some modern vitalists, the chief characteristic of living material, the one most difficult to explain chemically or physically, is the extremely complex organization of living systems.

The distinguished general physiologist Lillie believes that living material has a "directiveness" quite different from the "randomness" of inanimate material.

The mechanist is impressed with the fact that because of our newer knowledge of viruses there is no longer any sharp and clear-cut distinction between living and non-living, and he also likes to argue that vital phenomena once thought highly mysterious have now been explained, to some extent at least, in physical and chemical terms.

To the hard-working physiologist, intent on obtaining at least a partway explanation of some physiological process in physiochemical terms, the validity of the mechanist point of view scarcely seems worth discussing.

He is not concerned with the ultimate truth sought by the philosopher. But, unfortunately, when some elderly biologist loses faith in physiochemical interpretation as an answer to all the ultimate problems of biology and leaves his laboratory to publish defeatist or vitalistic views in philosophical journals, these views are promptly read by the philosophers, who by and large have very little acquaintance with biological journals and biological thought and who commonly are convinced that the vast majority of biologists are vitalists. Nothing could be farther from the truth.In view of the fact that general physiology is a new science, it has never been organized in the same way that older sciences have.

There has been no exact standardization of the scope of the subject and still less is there any agreement as to how it can best be approached, or how subdivided. Anyone may choose his own way of presenting the subject. After a brief survey of the morphology of living cells there will be a discussion of their chemical make-up in so far as it is known, and this will be followed by a consideration of the physical properties of living systems.

This preliminary survey of the statics of living systems will precede a more lengthy treatment of the kinetics. In considering the activities of the living machine the first section will deal with the chemical aspect of the subject; in other words, the living metabolism. This will include such topics as nutrition, digestion, respiration, secretion and excretion. The physical aspects of vital activity will then be taken up, and the energy output of the living engine—its production of mechanical energy, heat, electricity and light—will be discussed.

The living machine can function only under certain conditions of the environment, and the effect of the environment will next be considered. Living material can adapt itself to prolonged changes in environment, and we must therefore concern ourselves with the question of acclimatization. An outstanding characteristic of living things is their power to react in special and peculiar ways to sudden changes of the environment, that is to say, to stimulation.

This characteristic is known as "irritability," and there is a large body of information concerning various phases of it. After discussing irritability we shall conclude with a brief survey of the phenomenon of aging or senescence, and a short discussion of reproduction.

PHYSIOLOGY OF DEVELOPMENT

The great difference between the two branches of physiology is this—that while adult physiology is dealing with stability, developmental physiology is dealing with change. In the former case, the general characters of the organism, in spite of alterations in detail, remain the same over long periods of time; and one of the most prominent features of the processes of adult life is that, as a whole, they constitute a self-regulating system, in which any excess in one direction automatically brings into play forces acting in the opposite direction.

But in developmental physiology, the organism is always passing from one phase into another, and even if some of the phases—like the tadpole stage of the frog—are of comparatively long duration, yet processes are always at work which at length will upset the self-regulative power of the system, and will make it change into something quite different. Besides these processes of normal development, however, others of the same general type may occur as the result of special circumstances.

The facts of regeneration, for instance, come under this head; and, as a matter of fact, some analysis of regeneration will be found to give the best starting-point for a discussion of developmental physiology in general.

Regeneration is not a very familiar fact to most of us, since it is present to a very slight degree in ourselves and in the animals we know best.

But worms or polyps, or indeed most of the lower animals—if they could think—would find regeneration the natural, normal state of affairs, and the absence of regeneration a sad and abnormal failure. In a sense they would be right. Regeneration does at least seem to be a fundamental and original property of life, lost late and for special reasons. If you throw crumpled paper into a freshwater ditch, or set traps consisting of a bottle with boiled earth-worm for bait, you will probably succeed in catching large numbers of various kinds of freshwater Planarians—little leaf-shaped creatures of carnivorous habits and gliding movements.

Choose out the brown and black species, which are the hardiest. Transfer one to moist blotting-paper, and with a sharp knife cut it clean across. The experiment sounds cruel; but in reality you will have artificially encouraged the multiplication of the species. Each of the bits, replaced in water, will regenerate what is lost and so become transformed into a complete worm. The same would have happened if you had cut the animal into half a dozen cross-pieces instead of two, and equally whether you cut it down the middle or across.

If you persisted, you would discover two main facts in the course of your experiments—that any piece of any shape, provided it was above a certain quite small size, is capable of regenerating into a whole animal; and that this happens equally whether the regenerating process are fed or no. The same would have been true if instead of a planarian you had cut up the little polyp Hydra, and almost the same with the smaller Annelid worms such as Lumbriculus.

The same also holds for the microscopic single-celled Protozoa, although here the operations are much more difficult owing to the animals' small size. It is also true for the very early stages of development in many kinds of animal. In a newt either of the two first cells produced in segmentation may, if artificially separated, give rise to a whole embryo; the same holds in many animals for any one of the first four cells, or for most fragments of the late segmentation and blastula stages. Thus regeneration of the extreme type—what we may call complete regenerative capacity—is found in the simplest types of animal and the earliest stages of development.

It must therefore be thought of, not as a special property developed to meet the rare contingency of losing a limb or being bitten in two, but as something natural to living things, and present unless circumstances forbid it. Further facts confirm this view.

In animals of a rather higher grade of organization, such as the more complicated worms, crustacea, some insects, and some molluscs, the power to regenerate is still present, but it is limited. No longer can the animal be cut in half, or into a number of little bits, without losing its capacity for life; but so long as certain central and essential organs remain at work, very considerable losses, such as those of a limb or a tail, can be replaced.

This degree of regenerative power continues even into the land vertebrates. Although the frogs and toads do not, the lower or tailed amphibia still possess it. Even a mature newt or axolotl can grow not merely a new toe or new foot, but can repair the loss of a whole hind limb and limb-girdle. This we may call organ-regeneration. In such animals, the early stages in development of an organ such as a limb are capable of complete regeneration. Thus, by cutting a tadpole's limb-buds so that a number of separated bits are left, the animal may be made to grow more than the normal complement of legs. The same limitation of regeneration is seen, as might be expected, during development also.

For instance, while the amphibian in its segmentation and blastula stages has an almost complete regenerative capacity, the tadpole can only regenerate single organs, and the frog can do no more than heal wounds.

Even in newts and salamanders regeneration is much more rapid in the tadpole than in the adult. Among reptilia, lizards are the only animals which possess even the power of organ-regeneration, and in them it is confined to the tail. When we look closely we find that here the original general power of regeneration inherent in living things has been retained in this organ for particular biological reasons, and has been combined with special adaptations to make it of greater value to the animal.

The lizard has the power of self-mutilation (or *autotomy* as it is often called). If you catch a common lizard by the tail you will find that after a moment or two of squirming, the tail, broken off short at the root, will be all that you have in your hand, and the rest of the lizard will be running off into safe hiding.

This is only made possible by a special structure of the vertebrae of the base of the tail; they have a crack on either side which penetrates almost to their centre, and the tailmuscles are so arranged that when one set of them contracts they pull at the further half of one of these vertebrae, and snap it off across the plane of weakness. The tail when broken off is, though doomed to death within a few hours, not yet dead; and it continues to execute the most violent wrigglings and jerkings for some time.

In Nature what occurs is apparently as follows.—When a bird or other enemy makes a pounce at the lizard, it will often only succeed in catching it by the tail. When this happens, the tail is broken off, and by its squirmings continues to keep the attention of the attacker while the lizard itself is making for safety.

The tail is lost and devoured, but the lizard grows a new one. If lizard-eating animals were rational they would probably confine themselves to tails, gathering the annual crop as we gather the annual crops of fruit off a fruit-tree.

In any case, the lizard often saves itself by sacrificing its tail (*how* often could be established by finding out how long tail-regeneration takes, and then by examining a number of lizards from one locality and noting how many were in process of growing new tails—an experiment well worth trying); and this desirable end has only become possible by adding the special mechanism for breaking off the tail to the primitive power for regeneration.

Crabs and many other Arthropods have this power of autotomy in their limbs, and some worms when handled break their whole body into fragments, each one of which will regenerate into a complete individual! One word about the power of regeneration in mammals and ourselves. Although we cannot grow new limbs, yet our wounds will heal, and this is still a real regenerative process; we may call it wound-healing or tissue-regeneration.

It is also to be remembered that in a certain sense the growth of such tissues as the epidermis is a continual regeneration—for something is continually being lost, and as continually being renewed. These last processes fall under the head of *normal* or *physiological regeneration*. A more striking example, perhaps, than the skin is to be seen in the antlers of deer—two masses of bone weighing several pounds which are shed and renewed each year. Two or three general principles emerge from these facts. The first—to recapitulate—is that regeneration is a primitive property of life. Most low organisms never stop growing, and the capacity for regeneration is clearly associated with the capacity for growth, since growth is necessary for many of the processes of regeneration.

Other changes that are at work in regeneration, however, have the character of rearrangements; for instance, in the formation of a whole newt out of one of the first two blastomeres of the newt's egg, the only process is one of regulation—of rearrangement of material which ought to have produced a half, so that it actually produces a whole of half size.

In the regeneration of a small piece of Planarian to form a whole worm, a certain amount of new growth takes place at the cut ends of the piece, but many of the changes involved are brought about by changes in the original piece. For this to happen, however, a certain degree of plasticity is needful—cells and tissues must be capable of altering their original character and adapting themselves to new conditions. This also is largely a property of actively growing organs; further, it appears to be impossible if tissue-differentiation has reached too high a pitch.

Thus the gradual limiting of regenerative capacity which we find as a general rule both in the development of the individual and in the progress of life as a whole, is due partly to increased differentiation, partly to the fact that the higher animals, instead of growing indefinitely, show a sharply marked stoppage of growth on reaching the adult state. Regeneration is thus essentially the restoring of typical form, structure, and function; we might call it a special case of regulation.

The fact that a piece of Planarian without a mouth can yet regenerate head and tail is obviously interesting. The piece's capital can be readily transformed into cash for the undertaking of new enterprises—in other words, the living framework of its body can be broken down into foodmaterials which can then be used up in new growth elsewhere. In the higher animals, the living capital is locked up in non-negotiable forms to a much greater extent, although not so completely as is often imagined. This power of drawing upon the living tissues in case of need leads on to another remarkable power of Planarian and other low types.

If an intact animal of this sort is starved, it can continue to exist for periods of weeks and months by thus liquidating its capital for its day-by-day expenses. Since it draws on its own tissues for its daily energy-needs, it therefore must grow smaller and smaller. Planarians can continue this economical process until they have diminished their size to below that at which they hatched from the egg. If they are given food again (at any stage in the proceedings) they will start growing once more, and recover their normal size and appearance. One very curious fact was early noted about Planarians thus reduced in size by starvation—namely, that their proportions gradually altered from those of the adult and became more like those of normal young worms.

This led to the idea that perhaps they not only *looked* young, but had really *become* young again—an idea which was easily put to experimental proof. Some species of Planarians reproduce mainly asexually by transverse division; after the worm has reached a certain length a new head is formed a little behind the centre of the body, and in front of this a split appears along which separation eventually takes place. A brood of worms from one such species was taken. When they had nearly reached full size, they were divided into two lots. One lot, fed regularly, continued to grow and divide in normal fashion. The other lot was starved until the average size of its members was reduced to about half; then it was fed until it had recovered its original size; then starved again—and so on repeatedly. The experiment was continued for a period of time in which the well-fed animals went through nineteen generations (a period which, though only occupying a few months for Planarians, would in human beings mean about six centuries —from Chaucer's time till to-day). During the whole of this period, none of the other lot reproduced at all, but all remained within the limits of size set for them by the experimenter, and, what is more, showed no signs whatever of age or of diminished vigour. In fact there is every reason to suppose that the experiment could have been continued indefinitely—in other words, that the individual Planarian, by proper treatment, can be made immortal.

The elixir of life was sought by the alchemists of medieval Europe for hundreds of years. These experiments show that it has at last been discovered; but, unfortunately, it is only effective with Planarians and some other animals, all of a low grade of organization. Before we go further into the principles involved in this matter, we must mention that there is another way in which animals may surmount starvation and other unfavourable conditions. If an ordinary hydroid polyp like Obelia be kept in the laboratory, the hydranths, or separate organized individuals of the colony, will (unless the water is well aerated artificially) show a curious series of changes.

Their mouths will close and their tentacles become stumpier; eventually they will lose almost all signs of their previous differentiated structure and become converted into a rounded or egg-shaped bag with a few little knobs marking where the tentacles had once been. They have lost differentiation—in other words, have undergone dedifferentiation. In Obelia the process is complicated by the fact that while differentiation is going on, the cells of which the hydranth is made up detach themselves from their neighbours and, leaving their fixed place in the tissues, migrate into the central cavity and down into the common stalk which connects all the hydranths. By this means the hydranth dwindles rapidly, and is reduced soon after dedifferentiation is complete, to a tiny knob in the bottom of its protective horny cup.

The animal has been resorbed by the stem. It is as if a house were to be unbuilt, by the bricks leaving their places in the walls and migrating into the garden. If, however, the hydranth is cut off at its base beforehand, the central cavity is soon filled up with detached cells, and after this, dedifferentiation alone occurs until the animal is reduced to a mere living bag tight packed with cells. These types of dedifferentiation are found in many low types of animals, such as various Coelenterates and Ascidians. Among higher forms partial dedifferentiation may be a normal mode of development. For example, when the tadpole metamorphoses into the frog, some of its tissues start to dedifferentiate, for instance, those of the gills and tail. The process is complicated by the fact that phagocytosis, or the devouring of undesirable materials by white blood-cells, here reaches a far greater pitch of perfection than in lower animals, and, as a result, as soon as the cells have reached a certain degree of dedifferentiation, they are attacked and removed by the white blood-cells.

The disappearance of the tadpole's tail is thus due more to the activity of the phagocytes than to the migration of the tail-cells themselves. Similar methods of resorption of tissues are found when, for instance, a tumour or a graft disappears in a human being. To return to regeneration, and to another of its aspects, it is found that the presence of one part is often necessary for the regeneration of another part. In Planarians, for instance, the mouth opens at the end of a sort of trunk, the muscular pharynx, which can be protruded from the centre of the underside. If a small piece of the flatworm is cut out by two transverse cuts, it depends on a number of circumstances whether a head shall be formed or not; for instance, a low temperature near freezing, or exposure to dilute alcohol or other poisons (the experiment is easy to carry out) will reduce the frequency of head-formation, or even suppress it altogether. If a head is produced, such a piece will always proceed to the formation of a pharynx. But if no head appears, the formation of a pharynx depends on the original position of the piece in the body.

If it was cut from in front of the original pharynx, it will often regenerate a pharynx, even in the absence of a head; but if it came from behind the original pharynx, a new pharynx will never be formed unless a head forms first. In other words, a new pharynx will be regenerated in the centre of a piece of Planarian provided that there exists at the front end of the piece a region which is normally anterior to a pharynx—whether this anterior region be an old part of the body or a new regenerated head. Thus, the head or front region of the body exerts some remarkable influence upon a distant region of the rest of the piece, making it construct an organ that it could not otherwise produce. It makes things make themselves. Exactly how this effect is produced we do not know.

Possibly it is through the agency of the nervous system. In any event, it is clearly a property of very great importance, and also one which seems to be of quite general occurrence. For instance, when the eye-stalk of a prawn is cut off, a new one like the old is usually regenerated. But when, in addition to the eye-stalk being amputated, the optic ganglion, or part of the brain to which run the nerves from the eye, is also cut out, the organ which is regenerated bears no resemblance to an eye and eye-stalk, but has precisely the form of part or all of the first antenna or feeler, the appendage which normally comes next behind the eye. One other fact gained from the study of regeneration needs to be considered before the problems of normal development.

If one of a pair of organs be removed, remarkable results often follow. For instance, in certain diseases, one kidney may have to be cut out. When this is done, the other begins to grow, and at the end of a few months weighs almost as much as the original pair together. This enlargement follows as a direct result of the increased demands made upon the functions of the surviving kidney, and is therefore known as compensatory functional overgrowth. There is no regeneration of the missing organ, but a compensatory growth of its mate, resulting, one may say, in a *regeneration of function*. This functional overgrowth may be found not only to compensate for the loss of one member of a pair of organs, but also when part of a single organ is removed, or when an extra demand is made upon an organ without removal of anything.

If, for instance, most of the liver or pancreas or reproductive organs are removed, the bits that are left will start to grow rapidly, until they are large enough to cope with the demands of the body. Or, as illustrative of the second case, the normal thyroid secretion is very rich in iodine. When the amount of iodine in the food and drink falls below a certain quantity, the demands of the body for more iodine-containing secretion react on the thyroid; this begins to grow, producing more total amount of secretion as an attempt at compensation for its poverty in iodine. It may grow to a relatively huge size, producing a great swelling, called a goitre, in the neck. If a little iodine is given to a patient with such a goitre, the demand ceases, the reverse process sets in, and the swelling is reduced.

A similar process takes place in muscles when extra demands are made upon them. Everybody knows how heavy exercise 'develops the muscles', and, after a preliminary period of loss of weight due to utilization of the reserves of fat, makes a man put on flesh.

The actual size of the muscles increases as the result of exercise. With this information we can now attempt some account of the normal development. It is found that the early development of a vertebrate such as a frog or newt falls, from our present standpoint, into three main periods. The first is essentially a period of the division and rearrangement of already existing materials, and takes the fertilized egg through segmentation and some way into germ-layer formation.

The second is mainly one of differentiation—visible structure appears, the organs are blocked out, and tissues become different each from the other. Finally, in the third, the two most important features are growth, and the welding of the differentiated parts into a working whole—by mutual influence of organs upon each other and by an adaptation of separate organs to the demands upon them through the effects of function upon growth. This last phase may be taken first, since it is in many ways the most familiar. Think of a young organism—a recently hatched tadpole, a puppy, a child. Although it has already a characteristic organization, yet this organization is still plastic within wide limits.

The final shape and size of its bones, its muscles, its sinews and tendons, its glands, its digestive tube, even of parts of its central nervous system, depend very largely upon the amount and the kind of use to which they are put during development. If one of a puppy's fore-limbs is tied up soon after birth so that it is never used, the dog will manage to get about very well with only three legs. But the leg which has not been used will be very different from the others.

The bones of its skeleton will grow almost to the normal length, but will never attain more than about half their normal thickness, nor will the arrangement of bony struts and stays at either end of the shaft (which in an ordinary bone are developed so as to meet both vertical and sideways strain in the mechanically most perfect way) be properly formed. The muscles of the leg, too, and their tendons will be very small. Not only this, but the nerve-cells of that region of the brain which controls the movements of the limb will be undersized. The power of bone to respond to the strains to which it is exposed, is of practical value.

If, for instance, one of the small bones or phalanges of a finger is crushed and has to be removed, a small piece of bone from some other region can (if it include some of the periost layer from which new bone-cells are produced) be grafted in to take its place. After a year or two the irregular splinter will have come to bear a marked, though not complete, resemblance to the bone for which it is doing duty. One of the most striking adaptations in the body is the fact that every tendon which attaches a muscle to a bone not only is of the proper size for the strain which the muscle can exert upon it, but is composed of parallel fibres which run precisely in the region of greatest stress.

How can this beautiful relation of mechanism to function be supposed to have originated by natural causes? Does not the purposeful adaptation, different in detail for each separate tendon, demand the intervention of some supernatural power? Research, however, has in great measure removed such difficulties; for it has shown that all the separate detailed adaptations are the direct consequence of one inherent property of the tissue of which tendons are made. If a piece is cut out from a long tendon, such as the Achilles tendon which passes from the muscle of the calf over the heel to be fastened to the sole of the foot, it will after a time regenerate.

Connective tissue grows out into the wound, the cells elongate and form fibres in the direction of the tendon, make attachment to the cut ends, and reproduce a structure essentially like that which was originally present. But if by one means or another the calf-muscle is put out of action (as when its nerve has been severed) then, although the gap will be filled by a new growth of connective tissue, the fibres will not become arranged in any particular direction, and no tendon-like structure will be formed.

Further, when in such an animal (the main experiments were carried out in rabbits) a piece of clean silk was healed across the middle of the wound, at right angles to the original tendon, and then, by winding it up very slowly on a screw, increasing tension was put on one end of it, it was found that fibres were formed in the new connective tissue, but parallel to the silk, and therefore useless for any restoration of normal function. From these facts (supported by many others) it is clear that tendon-fibres are formed from connective tissue when the connective tissue is frequently stretched, and further, that they are formed parallel to the stretching force.

This stretching force is usually the force of muscular contraction (in growing animals there will also be found stretching due to growth of bones); so that tendons will be formed from ordinary connective tissue when muscles begin to contract, and will be 'adapted' to the direction and amount of stretching exerted by the muscle. In any case, all the apparent design seen in the adaptation of each several tendon to the demands made upon it, is apparently a direct result of this one reaction of connective tissue to tension.

To sum up, we may say that certainly most of the tissues of the body—possibly all—respond either throughout life or at least during growth (while their cells are still capable of multiplication) to the demands made upon them, and respond on the whole very advantageously from the point of view of the whole animal.

The skeleton, the muscles, tendons, many and probably all glands, the central nervous system—all these we have already seen to show functional adaptation. So does the skin. Everybody knows that (after the blister stage has been negotiated) continual rubbing will make the skin thicken. 'Horny-handed' is so hackneyed an epithet for manual labourers as to have almost become mere journalese, and every oarsman demonstrates the fact in his own person. So, too, the circulatory system responds to functional demands. If Everest is ever climbed, it will probably be because men can become 'acclimatized' to high altitudes and consequent low oxygen supply, largely, it would seem, through an extra production of red bloodcorpuscles by the bone-marrow and in consequence a greater capacity of the blood for capturing and transporting what oxygen there is. Furthermore, the size of a blood-vessel, both diameter of cavity and thickness of wall, is determined very largely-perhaps wholly—by the amount of blood which is pumped through it, and the pressure at which it is pumped.

Cases are known in which large arteries become blocked up. If the blocking-up process is not too rapid, one or more small branches of the vessel will gradually enlarge, and produce channels capable of transporting the requisite amount of blood without difficulty. This has been known to happen even when it is the human aorta, the main and indeed the only large artery springing from the arterial side of the heart, which has been blocked.

The same kind of adaptation is also true of the protective reactions of immunity. It is a familiar enough fact that children who have had measles once do not generally catch it again, but this is only one among innumerable instances, such as the over-production of diphtheria antitoxin by horses injected with small doses of the toxin or poison produced by diphtheria germs, which has reduced the disease from a cruel scourge to a serious unpleasantness, or the benefits given by vaccination.

All these again depend on one single property of higher animals; when unaccustomed proteins enter their system, they can (if the foreign substances are not in too great amount, and provided certain other conditions are fulfilled) not only destroy or inactivate them, but in the process generate an excess of the destroying substance or *antibody*, which remains for a longer or shorter time so that a second invasion can be more rapidly and effectively crushed.

Although many of these adaptations are possible throughout life, others can only occur during development, and most of them are essential for normal development, since only through them do the size and structure of the various organs come to correspond to their function and to the physical and chemical stresses imposed on them.

There is, however, another type of mechanism which is active through the later periods of development and plays an important role not only in this functional adaptation but also in fusing the parts of the organism into a more unified whole, and that is the endocrine system or sum total of the ductless glands. Many of them, such as the thyroid and the pituitary, have indispensable functions as regards proper growth. The thyroid regulates the rate of metabolism, increased amount of thyroid secretion producing an increase of oxidation in the body. As might be expected, this function seems to be of more importance to the well-being of animals such as mammals, with a normal high metabolism (great heatproduction and constant high temperature), than to those like fish or frogs, which do not have to be constantly generating large quantities of heat to keep their temperature above that of their surroundings.

Within the mammals, the absence or failure of the thyroid makes more difference to an adult man than to an adult sheep: the thyroidless sheep is but slightly abnormal, while the thyroidless man becomes notably fat, sluggish, and slow-minded. But its absence in the child, in whom growth and activity should be much greater, makes far more difference again than in the grown-up man; the thyroidless human child or *crétin* usually dies early, can only grow into a stunted dwarf, and never develops normal intelligence, the brain of man apparently requiring a high level of metabolism for its unfolding.

In Amphibia, metamorphosis from larva (tadpole) to adult is under the control of the thyroid. Tadpoles with their thyroid cut out refuse to be transformed into frogs, and grow to giant size in the water, while if a dose of thyroid is given to quite young tadpoles, they will metamorphose into froglets no larger than flies. The changes which take place in body and mind at puberty are also under the control of the endocrine system. The immediate control is exerted by the gonad itself.

It is important to find that a whole set of separate changes can be brought about by a single change in the ductless glands, and further, that all these changes are in the long run related to one function, that of reproduction. That a single original alteration in the chemical composition of the blood, an alteration appearing inevitably at the end of a definite span of growth, should bring about an increased growth-rate, the formation of mature spermatozoa in the testes, the appearance of hair on the face and body, and a marked change in mental outlook and temperament, is one of the most striking facts in biology.

All the ductless glands have this in common, that their secretions are carried all over the body by the blood, so that they can exercise a number of different effects simultaneously upon the most diverse organs. It is fairly obvious, however, that for all the importance of function and of the endocrine system in development, their task is essentially one of regulating the degree of activity of processes already in being, or of directing them along particular channels. Bones exist in the absence of a pituitary or of pressure upon them, and muscles without being used.

It is only a small part of the adult man's growth of hair that is dependent upon the testis; metabolism continues in the absence of a thyroid, and the first development of a gland or blood-vessel takes place before its use begins.

Another point that should not be forgotten is that in the long run the differences in shape and size between different animals depend almost entirely upon differences in the amount and direction of growth in different regions during development. In very early life the hind-limbs of a whale grow much less fast, compared with the body, than do those of a man, and those of a man much less fast than those of a kangaroo.

Some idea of the changes which would be produced in human beings if the proportions between general growth in length and growth in breadth were rather different from what they actually are, can be obtained by visiting the distorting mirrors at a fair or circus.

Or one organ may be affected differently from the rest; the effect of this on general proportions may be seen by looking at oneself in the broadening mirror, first with the arms held against the sides, and then outspread at right angles to the body.

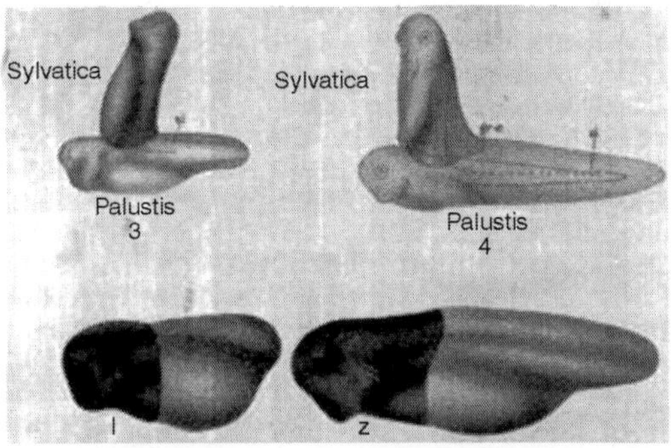

Fig. ***Two Experiments in Grafting in Tadpoles.***

- The front half of an embryo of one American species of frog (*Rana sylvatica*) has been cut off and grafted on to the hind half of another, lighter-coloured species (*R. palustris*).
- The compound animal (chimaera) grows quite normally. The lateral line, which originates near the head, has grown down from the *sylvatica* component on to the *palustris* trunk. Compound individuals like this have been reared through metamorphosis. 3 and 4. The front part of a *sylvatica* embryo has been grafted on to the back of a *palustris* embryo: this combination also has continued development. The *sylvatica* lateral

line (below *si*), on reaching the *palustris* component, has bent round into correct position.

If in the embryo of the fish Argyropelecus the tendency to growth in length had been a little less, and the main direction of dorso-ventral growth were not quite at right angles to the main direction of growth in length, it would have developed into a passable imitation of another kind of fish. No doubt such slight alterations in growth are at the bottom of much evolutionary change. How do these more fundamental structures and functions originate to provide the primal basis to be shaped by the secondary moulding forces we have described? Although much is still not clear, yet the main outlines of the answer emerge when we look at the next previous stage of development, the stage of differentiation. This starts during gastrulation, and ends as the various organs become capable of exercising their functions.

Its great characteristic is the sudden appearance, in the hitherto comparatively simple germ, of a number of chemically different regions, each one of which is irrevocably destined to generate one particular kind of organ. What is more, this will usually be accomplished whether the developing organ-rudiment is in its proper relation to the rest of the organism or not. If the late gastrula is cut in two so as to separate front from hind region, each part (after healing the wound) will proceed to develop just those organs which it would have done if the embryo had been left intact. We are presented with an anterior half-embryo without any of the hinder organs, and a posterior half-embryo with tail, hind limb-buds, and most of the trunk, but no head-regions, gills, or heart.

There is, in fact, in this stage *no* regeneration whatever of missing parts; yet, in spite of this, both halves appear quite healthy, and only begin to degenerate at a stage when the organs should be beginning to work and the animal to fend for itself. Or a piece of one embryo may be grafted on to another embryo, even of a different species, and both will continue to develop. One actual 'chimaera' has been artificially produced consisting of the front half of one kind of frog and the posterior half of another, which grew and metamorphosed quite happily. Or if a small part of the future brain-region be cut out just after the gastrula stage, and grafted in again in reversed position, it will produce what it would have produced in an untouched embryo, although now of course the organs to which it gives rise will be misplaced.

The eye, for instance, which is formed by an outgrowth from the brain, may by this means be made to develop behind the ear, and yet have normal structure. This will happen in spite of the fact that no visible trace of even the earliest eye-rudiment was present when the operation was performed. Before gastrulation, on the other hand, no such irreversible changes, inevitably determining the future appearance of organs, have yet set in. Cutting the developing egg in half in the earliest stages has a totally different result from cutting it in half after gastrulation has set in.

In the former case, as we have seen, one or both halves may become reorganized into a whole, and produce an embryo normal in every respect save size. One very remarkable fact has recently been discovered, namely, that the sudden change that takes place in a few hours during gastrulation in an animal like a frog or newt, during which the germ loses its capacity for regulation and becomes a sort of jig-saw of separate parts, each predetermined to produce just one particular organ of the future tadpole, occurs under the influence of one small region of the embryo.

This is the region just above the dorsal lip of the blastopore. By ingenious microscopic operations, even tiny bits of a newt's egg can be removed and grafted into other places on the same or other embryos. If this dorsal-lip region be taken and grafted into the flank region of another developing egg, it is found to cause the development of an extra set of organs, in addition to the one produced in relation to the host egg's own dorsal lip. An extra nerve-tube, notochord, muscle segments, kidney tubes, and gut may be thus induced to form—in other words, the beginnings of a whole second embryo.

What is more, they are not formed by the grafted piece itself, but in the host-tissues near it. Again, we are reminded of Mother Carey in Kingsley *Water Babies*, who 'made things make themselves'. No other region has this power, and so the title of 'organizer' has been bestowed upon the dorsal-lip zone. It is interesting to recall that a similar sort of power was exerted by the head of a Planarian in inducing the formation of a mouth and pharynx.

Another fact is that the dorsal-lip region at this period is the most actively working part of the embryo, with very rapid cell-division. It seems probable that this great activity has something to do with its organizing capacity. Other interesting results, also apparently concerned with differences of activity in different parts of the developing egg, have been obtained with still earlier stages.

Although the period up till early gastrulation is of course an essential link in the developmental chain, no important steps in the construction of the details of the animal's typical organization are being taken so long as it lasts. The original comparatively simple organization of the zygote is not much altered; and the most essential change is the cutting up of the huge and unwieldy single cell, the fertilized ovum, into hundreds of small cells.

4

Temperature Effect of Animals Body

The temperature of an animal depends on the amount of heat that is contained within the animals' body. Body heat is generally defined as the amount of heat produced by the body through metabolism added to any thermal exchange between the animal and the environment. There are several ways that an animal can absorb or lose heat from or to the environment: conduction, convection, radiation, and evapouration. Conduction is the transfer of heat between objects and substances that are in contact with each other. The transfer of heat from the skin surface to molecules of gas in the air is an example of conduction. Convection is the transfer of heat contained in a mass of a gas or liquid, by the movement of that mass. A simple example of this is when you sit in front of a fan on a hot day and the movement of air takes the heat away. Radiation is the transfer of heat through electromagnetic radiation without objects coming into contact.

An example of radiation is the warming that can occur to animals in sunlight. Evapouration is a change in state from liquid to gas, which requires energy. Heat is dissipated from the body during evapouration because large water molecules with high energy content accumulate at the skin surface and enter the gaseous phase. When the larger molecules vapourize, they take thermal energy with them and any water left behind becomes cooler. Animals have the ability to gain and store heat from the environment which can result in an increase in body temperature. The amount of heat that will transfer into or out of an animal depends on surface area, temperature difference, and the specific heat conductance of an animals body surface. A large animal (elephant) has a small surface-to-mass ratio and will heat up more slowly in response to environmental heat, whereas an animal with a larger surface-to-mass ratio will heat up faster. If the temperature difference between the environment and an animal's body is small, very little heat will flow into or out of its body.

Insulators like fur, feathers, and blubber have a very low thermal conductivity and help to reduce the rate of heat flow. In addition to these thermal regulating characteristics, animals can also regulate the exchange of heat between themselves and the environment using behavioural control and/or autonomic control. Behavioural control is used by animals when they move to a part of the environment where heat exchange favours the attainment of optimal body temperature. An example of this is when snakes or other reptiles choose to bask in the sun during daylight hours, but may avoid basking during midday, when the suns rays are most intense.

Autonomic control is being used when the temperature of the skin and extremities is controlled by the amount of blood flow to these areas of the body. Constriction of arterioles leading to the skin keeps warm blood from reaching the colder extremities and conserves heat at the body core. Autonomic control can also be used to absorb heat by increasing the amount of blood that flows to the surface tissues. Near the skin surface, blood will approach temperature equilibrium with the environment. The insulating properties of the blubber layer in whales is a good example of how autonomic control affects body temperature regulation.

Whales at the surface of the ocean will increase the flow of blood to their blubber layer to absorb radiant heat from the sun. When diving into colder depths, whales constrict their blood vessels and decrease the blood flow to the blubber layer so that heat is conserved at the core and not lost at body surfaces. Acclimatization is also a method of regulating heat exchange, but it involves changes in insulation quality or heat dissipation effectiveness due to long term exposure to either high or low temperatures.

TEMPERATURE CHANGES AFFECT

Temperature variation can cause changes at the molecular level. Homeoviscous membrane adaptations occur as a result of temperature changes. The cell membrane and many functions associated with the cell membrane are sensitive to temperature change. Low temperatures cause the membrane to enter a gel-like phase with very high membrane viscosity. High temperatures, on the other hand, cause the membrane to enter a hyperfluid phase with very low viscosity. Initially, acute temperature changes cause changes in membrane polarity and fluidity. On the other hand, after acclimation to both high and low temperatures, the resulting homeoviscous adaptations cause similar lipid polarization and membrane viscosity at both temperatures.

Temperature variation causes certain changes in enzymes. The rate of enzymatically controlled reactions changes due to temperature. This process is called enzyme acclimation. Enzyme acclimation can occur due to changes that occur in the molecular structure of one or more enzymes, changes in the quantity of the enzyme, or other factors that affect enzyme kinetics (for example nerve conduction). Ectotherms are animals that produce metabolic heat at low rates and rely primarily on thermal conditions of their surroundings. They maintain body temperature that is often equal or dependent on the temperature of the surrounding environment, and tend to have lower metabolic rates at low temperatures and higher metabolic rates at higher temperatures. Ectotherms that normally live in cool temperatures have a low body temperature, maintain adequate metabolism at very low levels of enzyme activity, and have enzymes that can function at low temperatures.

Ectotherms that normally live in warm temperatures function at very low levels of metabolism and rely heavily on behavioural regulation to control their body temperature. Certain reptiles also use autonomic control to regulate body temperature. For example the Galapagos marine iguana regulates its heart rate and blood flow during dives into cold water to reduce the amount of heat lost at its surface tissue (skin). It is often more difficult for ectotherms to survive in extreme conditions of cold or hot environments. Extremely hot conditions are not frequented by ectotherms.

Most ectotherms never experience such extremes due to the existence of the critical thermal maximum, a temperature above which long-term survival is not possible. Freezing and cold temperatures are dangerous to all living organisms because if ice crystals begin to form within a cell, the crystals will continue to grow in size as freezing progresses until eventually causing the cell to rupture. No animal is known to survive complete freezing of its tissues, but some ectotherms have come close. Freeze tolerance mechanisms are used by some ectotherms that actually freeze and remain alive. Some species of beetles can withstand freezing temperatures because they have extracellular fluid that contains a substance that accelerates crystal formation. As a result of more rapid crystal formation outside of the cells, the solute concentration inside the cell increases, lowering the intercellular freezing point. Red blood cells, yeast, and sperm can withstand freezing temperatures using this mechanism, provided that intercellular ion concentrations do not get high enough to cause damage to the cell organelles.

Freeze avoidance strategies are used by other ectotherms to prevent freezing. Freeze avoidance is usually characterized by supercooling or the presence of antifreeze substances. Supercooling is observed in fish that live in water temperatures below the freezing point of their bodily fluids. They remain unfrozen because ice crystals are unable to form due to lack of nuclei which are needed to initiate crystallization. Antifreeze substances are characterized by their capacity to lower the freezing point of body fluids. Glycerol, which is present in various species of arthropods and insects, is an example of an antifreeze that has been shown to lower the freezing point to as low as -47°C. Heterotherms are animals capable of varying degrees of endothermic heat production, but they generally do not regulate body temperature within as narrow a range as endotherms. Heterotherms are broadly broken into two categories, temporal and regional heterotherms. Temporal heterotherms are a broad category of animals whose temperatures vary widely over time and may exhibit daily temperature fluctuations. Camels behave like temporal heterotherms because they allow their body temperatures to fluctuate throughout the day - absorbing heat during the day, and releasing it at night. Regional heterotherms are characterized by the fact that they can achieve high core temperatures, while peripheral tissues and extremities approach ambient temperatures.

The most common form of regional heterothermy is found among animals that are able to conserve heat generated by muscle activity. Certain species of flying insects require their flight muscles to be pre-warmed to a specific temperature before they are able to fly successfully. During flight, these insects can achieve relatively high body temperature, while at rest their body temperatures will reach equilibrium with ambient temperatures. Certain fish, like tuna and some sharks are referred to as warm-bodied heterotherms because of their ability to maintain relatively constant, high core temperatures. The ability of these fish to attain such temperatures is dependent on the organization of the vascular system. Blood vessels in ectothermic fish are located centrally, while heterothermic fish have blood vessels located under the skin. The location of the blood vessels in heterothermic fish allows for countercurrent heat exchange to be utilized, minimizing the loss of heat. The location of swimming muscles in heterothermic fish also influences the retention of body heat. These muscles are located near the central part of the fish and since most of these fish do not stop swimming, the muscles do not cool down. Other types of fish have specialized thermogenic structures that maintain a specific heat in a certain region of the body. An example of this is the swordfish which is able to maintain the temperature of its brain and eyes about 10 degrees higher than its surroundings.

Ainmal Endotherms

One trait that distinguishes the major groups of vertebrates from one another is the way in which they control their body temperature. Most fish, amphibians, and reptiles have body temperatures that are close to the temperature of their environment. Fish, amphibians, and reptiles are ectotherms. An ectotherm is an animal whose body does not produce much internal heat. Its body temperature changes depending on the temperature of the environment. For example, when a turtle is lying in the sun on a riverbank, it has a higher body temperature than when it is swimming in a cool river. Ectotherms are sometimes called "cold-blooded," but that is misleading. The blood of ectotherms is often quite warm.

In contrast to the turtle, a beaver would have the same body temperature whether it was in cool water or on warm land. The beaver is an endotherm. An endotherm is an animal whose body controls and regulates its temperature by controlling the internal heat it produces. An endotherm's body temperature usually does not change much, even when the temperature of its environment changes. Mammals and birds are endotherms. They have a stable body temperature that is typically much warmer than their environment. Endotherms are often called "warm-blooded" animals.

Endotherms

Endotherms are animals that generate their own body heat and are able to maintain internal temperatures that are often stable within a few degrees (between 37°C – 38°C in mammals and about 40°C for birds). This is achieved through elevated basal metabolism in conjunction with heat-conserving and heat dissipating mechanisms. At stable, moderate temperatures regulation is maintained by enough metabolic activity to exactly balance the heat that is lost to the environment. The range of temperatures in which an endotherm can maintain the basal rate of heat production is called the thermal neutral zone. As the temperature drops below the endotherms' lower critical temperature (LCT), the basal metabolic rate is no longer sufficient to balance heat loss to the environment.

The endotherm must increase heat production by means of thermogenesis, which is the production of heat through either shivering or non-shivering methods. Heat production will increase linearly as temperatures continue to decrease so that colder temperatures yield greater rates of metabolic regulation to produce more heat. Eventually if the temperature drops too much, metabolic systems fail to compensate and the animal will die.

Endotherms are able to survive in cold environments through different mechanisms of adaptation. Fur or feathers are useful in reducing convection and the loss of body heat at the body surface of different animals that have fur or feathers. Thick layers of insulation can also reduce loss of heat. For example, blubber (fatty tissue found under the skin of whales) is a good insulator because it has a lower thermal conductivity than water, is metabolically inactive, and requires little profusion of blood. Endotherms are generally larger animals than ectotherms and this quality is useful in extremely cold environments because large animals generally have a lower specific heat conductance and relatively smaller surface areas which will reduce surface heat loss.

Countercurrent heat exchange is also utilized by endotherms to drastically reduce heat loss from appendages. As environmental temperatures rise above the endotherms' internal temperature, metabolic rate decreases with increasing temperatures.

When external temperatures exceed the thermal neutral zone the endotherm must be able to release, or dissipate excess body heat through passive or active heat-dissipating mechanisms, such as evapourative cooling by sweating or panting. All living animals are continuously producing heat at a basal level. If the body temperature can not be controlled and the heat dissipated, it is possible for the animal to overheat and die.

BIOTECHNOLOGY AND MONOGASTRIC NUTRITION

Most developing countries are experiencing difficulties providing sufficient food (staples) for their resident population. In general in most third world countries food production is increasing at a slower rate than the population growth rate. Developing countries on average are increasing food imports by 10% per year and this is likely to continue into the foreseeable future. Whilst ruminants need not compete with humans for food there are few economic, large scale systems for pig and poultry production established in developing countries which do not rely on grain based concentrate or on crops such as potatoes and other root crops which also take up land that could potentially be used for human food production. The recent developments of sugar cane as a source of feed (juice) for pigs, however, has a major advantage which resides in the high biomass production per hectare and a great potential to form the basis of an integrated farming system. There has been far too little emphasis on developing systems of production of pigs and poultry on agro-industrial by-products. The exception being the molasses and swill-feeding systems for pigs that have developed in Cuba.

By-products high in sugar such as molasses, palm oil sludge and fruit and vegetable pulps lend themselves to feeding to monogastrics. The constraints to production are a low protein content and in some cases an associated high level of fibre which is an intestinally indigestible carbohydrate. Pigs are able to utilise some fibrous feed, possibly up to 30% of their digestible feed intake because of caecal fermentation. It appears desirable that in the future, pig production should utilise at least a proportion of fibrous carbohydrate polymers. Free-range pig production, where breeding stock obtain a fair proportion of their feed from pastures is a good beginning forced on producers in Europe by the high cost of grain feeding in intensive fattening systems. If the microbial ecosystem in the caecum of the pig is well developed, basal diets high in sugar and low in fibre combine well with vegetable proteins such as cottonseed meal which are fibrous.

Potential Areas for Biotechnology in Pig Nutrition

Overcoming Feed Intake Problems on High Fibre Protein Supplements

The use of oil-seed protein residues for pigs and poultry are generally limited by the fibrous components and in some cases the presence of secondary plant compounds which may be toxic or simply lower production (*e.g.* gossypol in cottonseed meal). To overcome the constraint created by too much fibre it may be possible to develop pre-treatment to protect the protein and then find mechanisms for decreasing the fibre (*e.g.* growing fungi, such as the white rot species on the materials to decrease fibre and increase protein) or find methods to hydrolyse the cellulose to simple sugars before feeding.

Chemical treatment of the feed to hydrolyze fibre to monocarbohydrates may also be potentially possible (*e.g.* with pressurised steam and sulphur dioxide). Delignification with say *Coprinus* fungi and treatment by pressure/steam of fibrous crop residues to produce simple sugars for inclusion in pig rations is a further possibility requiring research. At least one group in the U.K. is attempting to introduce cellulase genes from *Clostridial* sources into embryos of pigs with the concept of producing pigs with a capacity to digest more cellulose. The enzyme, if expressed will be controlled by a promoter sequence encoding only in the pancreas this is given in a little more detail below as an example of:

- The possibilities,
- The problems and
- The requirements for basic research prior to any developments which might be applied.

Digestive Function Through Development of Transgenic Animals

By virtue of their hindgut fermentation, pigs have about the same capacity as humans for digesting plant structural polysaccharides. In theory, substantially more energy would be available to pigs (and other simple animals) if cellulose and hemicellulose could be digested in the small intestine and the sugars absorbed. In addition the complex carbohydrates of grains (*e.g.* β-glucans of barley), fibrous protein sources (*e.g.* cotton seed meal) could be more efficiently used. In the latter case it would allow a much higher proportion of protein from such sources to be incorporated into the diets replacing more expensive less fibrous protein meals such as fish or meat meal. The ability to digest cellulose and hemicellulose by intestinal enzymes is not present in any animal. It is probable that in early evolution these abilities were not advantageous.

However, if encoding a gene for microbial cellulase or hemicellulase could be incorporated into an animal's genome and expressed in the pancreas it may allow these enzymes to be produced and secreted along with the other pancreatic secretions. This is being tested by Hazelwood and Gilbert by introducing into the early embryo of mice (and they suggest they will do this with pigs) the cel E gene of *Clostridium thermocellum* which encodes a thermostable endogluconase with xylan-hydrolysing activity. Expressions of this gene in transgenic mice will be regulated by the elastase I promoter/ enhancer region (from rats) which is located "upstream" of the elastase gene and is responsible for restricting the synthesis of the digestive enzyme elastase to target sites of expression of other genes, to the exocrine enhancer cells in the pancreas of transgenic mice.

To modify porcine digestive enzyme secretion to include the enzymes, cellulase and hemicellulase, will require tissue specific expression of the cel E gene and secretion of mature active endogluconase enzyme into the intestinal lumen. For this reason gene constructs incorporating the elastase promoter/ enhancer and cel E coding sequence have been designed. The concept of transgenic pigs is at the ideas stage, the research will undoubtedly require a return to basic investigations, undoubtedly over and under expression of the genes when transferred will be a major problem. Fibre breakdown in the pig by cellulase is likely to be slow and therefore gut capacity is likely to be a major constraint even when transgenic pigs are produced.

Porcine Growth Hormone (PSt.)

Undoubtedly porcine growth hormone injected in controlled amounts into pigs on grain based concentrates increase their growth rate, efficiency of growth and reduce fat deposition.

The place of PSt in pig production, as it pertains to developing countries, is unknown. The small production units, the likely high cost of injections and the "scavenger system" generally operated at the small farmer level in developing countries suggests that PSt is unlikely to be used. The exception may be in grain-based feeding systems for pigs aimed at the markets provided by the higher income groups and tourists.

Transgenic Pigs—Porcine Growth Hormone

The technology for production of transgenic pigs has certainly been developed. Problems of over and under expression of the genes are still apparent. Overexpression results in considerable problems of leg weakness and reproductive inefficiencies whereas under expression gives no additional benefits.

The nutrition of pigs expressing for growth hormone again has been under-studied with the usually emphasis being placed on the thrust to produce transgenic animals. The need to develop special feeding and management systems for pigs seems to have been ignored. Where these techniques have been applied in developed laboratories the absence of disease is mandatory. It is conceivable that in addition to a "special type of nutrition" there may be special requirements for disease and parasite control. Disease and parasitism are more likely to impact heavily on pigs and other animals with a greater potential for growth.

The problems to be faced in pigs transgenic for porcine growth hormone are well summed up in a paper by Wilmut et al.: "The pig exhibiting human growth hormone did not grow any faster than normal pigs but they had less back fat which might be of potential value to pig farmers. Unfortunately, these transgenic pigs also suffered from a number of abnormalities, including arthritis, lack of co-ordination of their rear legs, susceptibility to stress, anoestrus in gilts and lack of libido in boars. So it is back to the bench to attempt to redesign the gene to avoid these problems."

BODY TEMPERATURE

The body must be kept at a constant temperature, within a small range, in order for all of the systems to work properly. This is the normal body temperature. A change in the temperature of the body is a sign of ill health.

All mammals are endothermic—they maintain and regulate their own body temperature. Living in widespread environments around the world, mammals face daily and seasonal fluctuations in temperatures and some—for example those living in harsh arctic or tropical habitats—face extreme cold or heat. To maintain their correct body temperature, mammals must be able to produce and conserve body heat in colder temperatures as well as dissipate exess body heat in warmer temperatures.

In a healthy individual, body temperature is kept constant in a very small range despite of big differences in temperature of the surroundings and also those in physical activity. Very perfect regulation of body temperature, necessary for optimal progress of enzymatic reactions, is developed in all homoiotermic animals. It doesn't apply to poikilothermic animals. During the most variable changes in human organism, the body temperature may increase. Fever is a natural reaction during a number of illnesses. In several cases, absence of the natural reaction is more alarming sign than the presence of fever itself. Fever is usually accompanied by different general symptoms, such as sweating, chills, sensation of cold, and other subjective sensations. Missing of these symptoms during high temperature may be a sign of a serious illness.

The main task in heat production has thermogenesis caused by the effect of thyroid hormones. Hormones of thyroid gland stimulate Na^{+}–$^{+}$ATP-ase found in cytoplasmic membranes. Increased production of heat is achieved by increasing the metabolic processes in which energy is released in the form of heat. The greatest importance is splitting of ATP when 54kJ are released from one mol of ATP. Skeletal muscles, liver, splanchnic organs, and brain are the biggest producers of heat in an organism. In the heat production the muscles have especially important role. Because of their weight, they are able to produce very large amount of heat very quickly. Increased production of heat takes place in skeletal muscles during increased physical activity.

During the digestion, an increased production of heat occurs also in the GIT. Constant body temperature is achieved by perfect nervous regulation. Nervous system maintains the optimal intensity of metabolism and at the same time regulates the amount of heat loss. In early postnatal development, the thermoregulation is inadequate because of immature CNS. Fever is always achieved by reset the centre of thermoregulation to higher values. Hyperthermia means overheating of an organism caused only by exogenic causes (*e.g.* hot environment, hot bath). In this case the centre of thermoregulation doesn't change its setting up.

Heat is lost from an organism in several ways. The biggest loss is by conduction. It depends on the gradient between the body temperature and the temperature of the surrounding environment. The second way is by radiation. The third way is by evaporation. It is used especially during increased production of heat. Distribution of heat is done by blood circulation. Heat goes from each cell to the surrounding liquid and afterwards to the circulated blood. Modulating factor of heat loss is the amount of blood that circulates through the body surface. The big flow through the subcutaneous area and the skin secures the income of heat that may be given to the environment through the body surface.

Sweating helps delivering the heat. Sweat glands are controlled by cholinergic impulses through the sympathetic fibres. During intensive sweating, up to one litre of sweat may be formed. When the humidity of the environment is higher, a loss of heat by sweating is easier. When it is necessary to accumulate the heat in an organism, adrenergic stimuli cause reduction of the blood flow through the skin. The skin becomes an isolator decreasing the heat loss to minimum. Control mechanisms regulate the production of heat and its loss. Production and handover (loss) of heat are controlled from the centre in the hypothalamus.

It works on the principle of negative feedback control and includes:

- Receptors registrating central temperature
- Effector mechanisms composed of vasomotors, metabolic effectors, and controls of sweat glands
- Structures recording whether the actual temperature is not too high or too low

Increased central temperature activates mechanisms enabling the heat loss. Low central temperature activates mechanisms enabling the accumulation of heat. These mechanisms work as the thermostat. In healthy individuals, the body temperature (oral temperature) is somewhere between 36.5°C and 37.5°C. It slightly increases during the day since the morning (from 6:00 a.m.). The peak is reached at 6:00 to 10:00 p.m. The lowest temperature is between 2:00 and 4:00 a.m. Diurnal variation depends on the activity throughout the day.

Diurnal variations don't change in persons that work at night and sleep during the day. Such a diurnal variation is also kept when fever occurs. Fever reaches the peak in the evening, and in the morning even a very sick patient may have almost normal temperature. Body temperature changes are more intensive in young person than in old people. The temperature may slightly or temporarily increase in hot environment. Physical activity may also increase the body temperature. In extreme effort, the increase may be very high. The temperature in marathon runners may increase to 39°C to 41°C. The temperature may increase slightly if vasodilatation, hyperventilation, and other compensation mechanisms fail. Small increase in temperature may occur if the surrounding temperature is lower or the jogging is done early in the morning.

Organism uses simple mechanism for temperature regulation. It is the blood flow through the skin and subcutaneous area. Vasoconstriction allows the increased accumulation of heat, and vasodilation secures its quick loss. Changes in temperature up to 3°C don't cause an interuption of physiological functions. Spasms may occur during high fever in children. If the body temperature is increased over 42.2°C, irreversible changes in the brain occur. In humans the temperature usually doesn't overcome 41.1°C. Uncontrolled decrease in temperature below 32.8°C is accompanied by confusions and gradual loss of consciousness. If the decrease continues under 30°C, the fibrilation of ventricles occur that is the sign of fatal termination of this condition.

Brown fat that differs from the white one in structure and sites of location has an important function in thermogenesis in newborns and children. It is found between scapulas, on the neck, in axils, around the aorta and the kidneys. It is highly vascularized, and it has large mitochondria in its cells. One could say that while the white fat acts as feather-bed, the brown one is an electrical pillow. Receptors of cold conduct the information to the centre of thermoregulation. From this centre the impulses run in the sympathetic nerve fibres and lead to the release of norepinephrine in the brown fat. Norepinephrine activates the enzyme lipase. Activated lipase splits the fat to glycerol and free fatty acids (FFA) and the heat is released. Glycerol and FFA remain in the cell and can perform the resynthesis after some time. An adult person has little brown fat.

HEAT PRODUCTION

The mechanisms mammals have for producing heat include shivering, cellular metabolism, and circulatory adaptations. At some point in our lives, we've all shivered. Shivering generates heat as muscles quickly contract and shake. Cellular metabolism is the chemical breakdown process that occurs within cells. This process releases heat and warms the body. Circulatory adaptations, such as countercurrent heat exchange introduced above, transfer heat from the core of the animal's body to the periphery by specially designed blood flow paths. Warmer blood from the interior of the animal flows to cold extremities, thereby moderating the temperature of the more exposed limb.

Warmth is one of the necessities of life. Vital activities are possible only between certain narrowly defined limits of temperature. Cold inhibits and excessive heat suspends them. Body heat is energy. It is employed not just in resisting cold, but also in accelerating cellular activities. Temperature, within certain narrow limits, is so absolutely essential to life that all functions are excited by any attempt at its variation. Animals are roughly divided into two major classes: warm-blooded and cold-blooded.

This is according to whether they have means of producing and maintaining their own temperature or are dependent upon the surrounding medium (water or air) to provide it. The invertebrates, although they breathe oxygen and circulate fluids throughout their bodies, have no red blood corpuscles and are cold-blooded animals. Fishes and reptiles, vertebrates with red blood cells, are also called cold-blooded animals, although they are able to maintain an internal temperature above that of the surrounding water or air. Invertebrates have no heat of their own, but receive their temperature from the surrounding media and adapt to it.

Except for fishes and reptiles, whose heat-producing and heat-regulating powers are very limited, we may say that all vertebrates are warm-blooded, having red corpuscles, while the reverse is true of the invertebrates which have no red corpuscles. It may be suggested that since animals can live without red corpuscles and exist without internal heat, the primary office of respiration is more universal than to provide for the production of animal heat. Using a popular phrase in biology, heat production is only a "secondary adaptation." If we look at a large number of lower animals, we find them to be small and living in water. This medium directly and powerfully reduces them to its own temperature, and they are surrounded and permeated with water. In the radiata water actually mingles in large quantities with their digested food, so that they must of necessity remain at or very near water temperature.

Even if they possessed sources of heat within themselves, it becomes evident that heat production cannot be the great end of respiration in these animals. Its primary function must be something very different from this. If we take a second look at these animals, we discover that large numbers of them, especially those that live in fresh water, vary in temperature with the medium in which they live. Often they vary to a great extent, being sometimes near the freezing point and at other times fifty to one hundred degrees above it. Although a particular temperature may be best for each of them, still, many of them can live an active life in temperatures seventy, sixty, fifty and even forty degrees less.

It is obvious that to a small extent they could raise the temperature of their bodies above that of the water, when it is forty or fifty degrees, would be of no great importance. In their case at least, there must be some more important end for respiration than production of heat. Heat supplies a necessary condition of vital activity. The activities of cold-blooded animals rise and fall as the temperature goes up or down. The higher the temperature, providing it does not go so high so as to destroy life, the greater the activity. If it becomes very cold, they suspend activity. It is not in inorganic chemistry alone that heat promotes the energy and intensity of action.

In "vital chemistry," that is, in living functions, the same phenomenon is observed. An elevation of temperature accelerates all vital functions, both in the cold-blooded and in the warm-blooded animals. A similar thing is seen in plants. Acceleration of activity increases with the rise in temperature until the temperature reaches a certain variable optimum, after which any added increase in temperature reduces activity. The rate of activity for some of the lower forms may become so great as the temperature rises that they "live too fast" and wear themselves out.

When temperature is lowered, vital activities are lowered. In the cold-blooded animals, some of which may be frozen for long periods and then revived, all activity ceases after the temperature is reduced below a certain variable minimum. Most of the warm-blooded animals die when frozen, their vital activities ceasing before they reach the state of freezing. Higher animals are not so dependent upon the surround-ing temperature. They are not only equipped with internal sources of heat and mechanisms to control its production and radiation, but they also in most instances have outer coats of hair, feathers or wool to protect them from the cold. They possess means of lowering heat production and increasing heat radiation if the external temperature or their own internal heat due to activity is increased. (By the operation of the same internal heat-regulating mechanism they produce fever when needed.)

Thus, while the very form and habits of the lower orders of life are determined by external surroundings, the forms and habits of life of the higher animals are very largely determined by powers within them. These often prevail over powerful antagonistic forces. The lower animals are more or less slaves to the external world; the higher animals make the external world serve them. It should be noted that this independence of the higher animals, this internal energy, is in great measure due to a capacity for maintaining their normal temperatures amid the changes in that of the surrounding water and air. The uniform temperatures maintained by higher animals promote and secure a constancy, precision and energy in the nutrition of their tissues, and in the vital functions that supply the animal with resources to carry on active life in the face of opposing influences in the world. A brief glance at the method of maintaining body temperature may be helpful. In respiration we learned of the office of oxidation in the production of heat.

It is necessary that we understand that the body is capable of both increasing and decreasing its rate of heat production as the external temperature falls or rises. These processes are rigidly controlled by the nervous system and fail only in greatly enervated and diseased organisms. But the body also increases and decreases the radiation of heat from the body as need arises. While oxidation warms the body, evaporation (as in sweating) cools it. These physiological processes are carried on in relation to vital wants. The human body, to narrow our considera-tions at this point, is based upon a system of self-regulation and equipoise, and its temperature relations are beautifully provided for. In a cool atmosphere less heat is lost by evaporation and more produced within the body, while a reverse process is seen in a warm temperature. In all changes of temperature outside the body, some compensatory effort is required. But if our other relations are correct, the internal heat-regulating capacity of the body will be efficient. The maintenance of the heat-making mechanism of the body is an indispensable condition of health. Feeble and sick individuals who find it difficult or impossible to maintain normal temperature in a cold climate need to be kept warm. Chilling inhibits all functions of life and reduces their already greatly reduced stock of energy. The escape to a warm climate is no mere luxury for such persons. Warmth of some degree is certainly a normal requisite of life. But experience and experiment have shown that when the temperature of the surroundings is out of all proportion to the needs of the body and to its capacity to adjust itself, the body must and does suffer. There is not only discomfort, which normally causes us to seek relief from extremes of heat or cold, but there is some expenditure of energy in resisting extreme temperature.

In lands where fogs, frost and darkness cramp the energies of man, as well as in regions where excessive and long-continued heat depresses his vital activities, life is handicapped. By means of clothing, housing and artificial heating arrangements, we are able to live in cold climates. By means of cooling systems and a reduction of clothing, we live more comfortably in hot regions and seasons. But none of these arrangements are ideal. A warm climate serves man best; first-class habits of living enable him to live better in whatever climate he resides.

Disipating Heat

In warmer climates, excess body heat can accumulate and cause life-threatening problems for a mammal. Circulation near the skin's surface releases heat into the environment. Moisture from sweat glands or respiratory surfaces evaporates and cools the animal. Unfortunately, evaporative cooling is less effective in dry climates where water loss is also costly for mammals. In such situations, mammals often seek cover during the hotter daylight hours and resume active at night.

The normal body temperature

The body can only work properly at a certain temperature. The animal body maintains itself at a constant temperature, within a small range, in order for the systems to work properly. This normal body temperature is different in different types of animals.

There are a number of ways by which animals control the temperature of the body:

- Hair, wool, walking, running, shivering and the burning of energy in feed keep the body warm.
- Sweating, panting, wallowing in mud, and lying in the shade cool the body.

MEASURING BODY TEMPERATURE

We use a thermometer to measure the temperature of the body. The unit of measurement is degrees centigrade (°C). The normal temperature of your body is 37°C. We measure the body temperature of animals by placing a thermometer in the anus.

The thermometer

- Look at your thermometer. Notice the silver line of the mercury inside it and the scale with numbers marked along it.
- Before you use it you must make sure that the mercury level is below 35°C. If it is not, shake the thermometer to make the level go down.
- Every time you use the thermometer clean it with cold water and soap or disinfect it afterwards.

Do not wash the thermometer in hot water as this will burst it. Do not leave your thermometer in the sun as this may burst it. Carry the thermometer in a case in your pocket or bag. Do not use your veterinary thermometer for people.

TAKING BODY TEMPERATURE OF ANIMALS

- Control the animal.
- Move the tail to the side.

- Put the thermometer gently into the anus, as far as possible.
- Hold the thermometer at an angle so that it touches the wall of the rectum. Keep a firm grip on the thermometer, if the animal defecates or coughs the thermometer could come out or go into the rectum.
- Hold the thermometer in place for half a minute. If you do not have a watch count slowly up to 30 (one, two, three............ thirty).
- Remove the thermometer and wipe it if necessary and read it. Do not touch the bulb as this could change the reading.

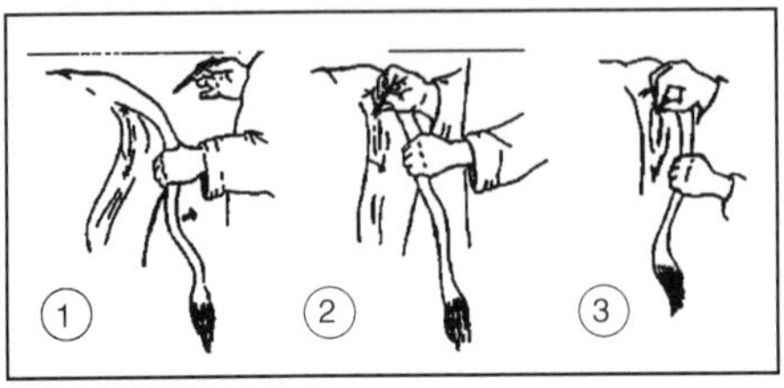

Normal body temperatures

Animal	Normal Temperature °C	Animal	Normal Temperature °C
Cattle	38.5	Calf	39.5
Buffalo	38.2	Goat	39.5
Sheep	39.0	Camel*	34.5-41.0
Llama, alpaca	38.0	Horse	38.0
Donkey	38.2	Pig	39.0
Chicken	42.0	Piglet	39.8

Body temperatures may be 1°C above or below these temperatures.

* The camel's body temperature will vary with the time of day and water availability. When a camel is watered daily its body temperature rises from 36.5°C in the morning to 39.5°C at noon, if the animal has no water, the temperature range is 34.5°C to 41°C. If you suspect that the animal has a high temperature use your thermometer to check it. Remember that a high temperature is one sign of ill health. When an animal has a high temperature it has a fever.

WARM-BLOODED AND COLD-BLOODED

Warm-blooded creatures, like mammals and birds, try to keep the inside of their bodies at a constant temperature. They do this by generating their own heat when they are in a cooler environment, and by cooling themselves when they are in a hotter environment. To generate heat, warm-blooded animals convert the food that they eat into energy. They have to eat a lot of food, compared with cold-blooded animals, to maintain a constant body temperature. Only a small amount of the food that a warm-blooded animal eats is converted into body mass. The rest is used to fuel a constant body temperature.

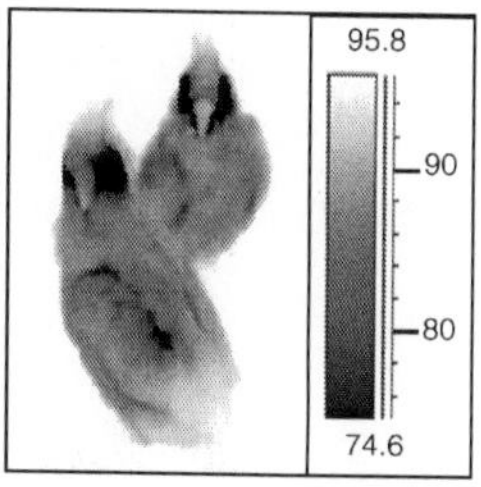

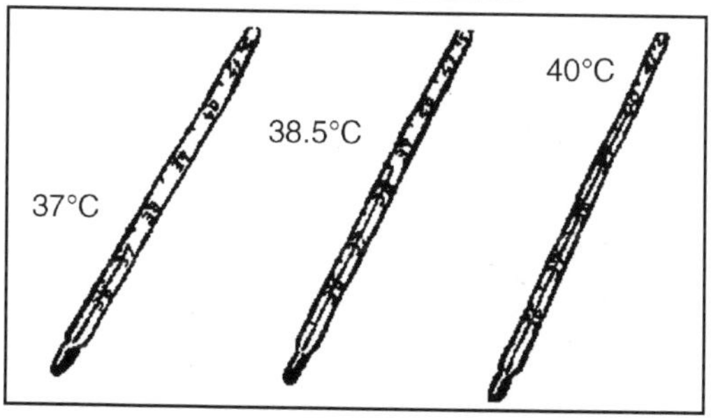

These thermal infrared images of warm-blooded animals, show how birds and mammals maintain body temperatures well above the surrounding, cooler air temperature. Cold-blooded creatures take on the temperature of their surroundings.

They are hot when their environment is hot and cold when their environment is cold. In hot environments, cold-blooded animals can have blood that is much warmer than warm-blooded animals. Cold-blooded animals are much more active in warm environments and are very sluggish in cold environments. This is because their muscle activity depends on chemical reactions which run quickly when it is hot and slowly when it is cold. A cold-blooded animal can convert much more of its food into body mass compared with a warm-blooded animal.

The infrared images above, show how cold-blooded animals take on the temperature of their surroundings. Both the gecko and the scorpion are at the same temperature as the air surrounding them. Notice the difference between these cold-blooded creatures and the warm-blooded humans holding them.

Staying Cool and Keeping Warm

To stay cool, warm-blooded animals sweat or pant to loose heat by water evaporation. They can also cool off by moving into a shaded area or by getting wet. Only mammals can sweat. Primates, such as humans, apes and monkey, have sweat glands all over their bodies. Dogs and cats have sweat glands only on their feet. Whales are mammals who have no sweat glands, but then since they live in the water, they don't really need them.

Large mammals can have difficultyin cooling down if they get overheated. This is why elephants, for example, have large, thin ears which loose heat quickly. Mammals have hair, fur or blubber, and birds have feathers to help keep them warm.

Many mammals have thick coats of fur which keep them warm in winter. They shed much of this fur in the summer to help them cool off and maintain their body temperature. Warm-blooded animals can also shiver to generate more heat when they get too cold. Some warm-blooded animals, especially birds, migrate from colder to warmer regions in the winter.

They will also expand their rib cage to increase their surface area and will darken their skin to absorb more heat. When a reptile is too hot, it will lie parallel to the sun's rays, go into a shady area, open its mouth wide, lighten its skin colour or burrow into cool soil. Some cold-blooded animals, such as bees or dragonflies, shiver to stay warm when in a cold environment.

Fish who live in areas where the winters are cold move to deeper waters during the colder months or migrate to warmer waters. Some fish have a special protein in their blood which acts like anti-freeze to help them survive very cold water temperatures. Snakes, lizards, toads, frogs, salamanders and most turtles will hibernate during cool winters. Some insects die when it gets too cold, however others survive by migrating to warmer areas or moving underground. Honeybees stay warm by crowding together and moving their wings to generate heat.

These thermal infrared images of a collard lizard show a cold-blooded animal's body temperature in a cooler and warmer environment. In the image to the above top, the lizard is just above room temperature, being warmed by the human hand holding it. To warm up lizards will seek a sunny area and bask in warm sunlight, as in the image to the above bottom.

Advantages and Disadvantages

There are many advantages to being warm-blooded. Warm-blooded animals can remain active in cold environments in which cold-blooded animals can hardly move. Warm-blooded animals can live in almost any surface environment on Earth, like in arctic regions or on high mountains where most cold-blooded animals would have difficulty surviving. Warm-blooded animals can remain active, seek food, and defend themselves in a wide range of outdoor temperatures. Cold-blooded animals can only do this when they are warm enough. A cold-blooded animal's level of activity depends upon the temperature of its surroundings.

A reptile, for example, will increase its body temperature before hunting and is better able to escape predators when it is warm. Cold-blooded animals also need to be warm and active to find a mate and reproduce. Being cold-blooded, however, also has its advantages. Cold-blooded animals require much less energy to survive than warm-blooded animals do. Mammals and birds require much more food and energy than do cold-blooded animals of the same weight. This is because in warm-blooded animals, the heat loss from their bodies is proportional to the surface area of their bodies, while the heat created by their bodies is proportional to their mass.

The ratio of a body's surface area to its mass is less the larger the animal is. This means that larger warm-blooded animals can generate more heat than they loose and more easily keep their body temperatures stable. Smaller warm-blooded animals loose heat more quickly. So, it is easier to stay warm by being larger. Warm-blooded animals cannot be too small or else they will loose heat faster than they can produce it.

Being large requires a greater food supply, but for mammals, being small also requires a lot of food to generate more heat. Small mammals need to eat very often to survive because they need more energy to keep a constant body temperature. They also need to eat high energy foods such as fruit, seed, and insects and even other small mammals. Larger mammals can get by with eating lower energy foods or eating less often. In an environment where food is scarce, such as in deserts, reptiles have an advantage. Since cold-blooded animals do not have to burn a lot of food to maintain a constant body temperature, they are more energy efficient and can survive longer periods without food. Many cold-blooded animals will try to keep their body temperatures as low as possible when food is scarce. Another disadvantage to being warm-blooded is that warm-blooded bodies provide an nice warm environment for viruses, bacteria and parasites to live in. Mammals and birds tend to have more problems with these infections than do reptiles, whose constantly changing body temperatures make life more difficult for these invaders.

However, an advantage of this is that mammals and birds have developed a stronger immune system than cold-blooded animals. A reptile's immune system is more efficient when the animals is warmer, however, since bacteria probably grow more slowly in lower temperatures, reptiles sometimes lower their body temperatures when they have an infection.

Exceptions

Some animals do not fall neatly into the categories of being warm or cold-blooded. Bats, for example, are mammals who cannot maintain a constant body temperature and cool off when they are not active. Echidnas maintain a range of body temperatures which usually lies between 77 and 98.6 degrees Fahrenheit and have difficulty cooling down. Mole Rats are another group of mammals who are unable to regulate their body temperature, however, since they live underground, the temperature of their environment does not change much.

Some warm-blooded animals, such as bears, groundhogs, gophers and bats hibernate during the cold winter. During hibernation these animals live off of stored body fat and can drop their body temperatures by as much as 50 degrees Fahrenheit. The Hawk Moth is an insect which can raise its body temperature well above the temperature of its surroundings when it is flying because of its huge wing muscles which generate heat when in use. Bees are another example of insects that can raise their body temperatures above that of their environment by moving their wings rapidly to generate heat.

TEMPERATURE IN VERTIBRATES

By numerous observations upon men and animals, John Hunter showed that the essential difference between the so-called warm-blooded and cold-blooded animals lies in the constancy of the temperature of the former, and the variability of the temperature of the latter. Those animals high in the scale of evolution, as birds and mammals, have a high temperature almost constant and independent of that of the surrounding air, whereas among the lower animals there is much variation of body temperature, dependent entirely on their surroundings. There are, however, certain mammals which are exceptions, being warm-blooded during the summer, but cold-blooded during the winter when they hibernate; such are the hedgehog, bat and dormouse. John Hunter suggested that two groups should be known as "animals of permanent heat at all atmospheres" and "animals of a heat variable with every atmosphere," but later Bergmann suggested that they should be known as "homeothermic" and "poikilothermic" animals. But it must be remembered there is no hard and fast line between the two groups.

The respiration and heart-rate being also retarded during this period, the resemblance to the condition of hibernation is considerable. Again, Sutherland Simpson has shown that during deep anaesthesia a warm-blooded animal tends to take the same temperature as that of its environment. He demonstrated that when a monkey is kept deeply anaesthetized with ether and is placed in a cold chamber, its temperature gradually falls, and that when it has reached a sufficiently low point (about 25 °C in the monkey), the employment of an anaesthetic is no longer necessary, the animal then being insensible to pain and incapable of being roused by any form of stimulus; it is, in fact, narcotized by cold, and is in a state of what may be called "artificial hibernation."

Once again this is explained by the fact that the heat-regulating mechanism has been interfered with. Similar results have been obtained from experiments on cats. These facts with many others tend to show that the power of maintaining a constant temperature has been a gradual development, as Darwin's theory of evolution suggests, and that anything that interferes with the due working of the higher nerve-centres puts the animal back again, for the time being, on to a lower plane of evolution.

HEAT PRODUCTION IN BIRDS AND MAMMALS

In cold environments, birds and mammals can compensate for heat loss by:

- Utilizing small smooth muscles (arrector pili in mammals) which are attached to feather or hair shafts; this shivering thermogenesis distorts the surface of the skin as the feather/hair shaft is made more erect (called goose flesh or pimples).
- Animals in cold climates tend to be larger (easier to maintain core body temperature) than similar species in warmer climates.
- Be capable of storing energy as fat for metabolism
- Have reduced extremities
- Some have countercurrent blood flow in extremities (*e.g.* timber wolves) to avoid freezing of tissues.

In warm environments, birds and mammals avoid overheating by:

- Behavioural adaptations like living in burrows during the day and being nocturnal
- Evaporative cooling by perspiration and panting
- Storing fat reserves in one place (*e.g.* camel's hump) to avoid its insulating effect
- Elongate, often vascularized extremities to conduct body heat to the air.

There are limits both of heat and cold that a warm-blooded animal can bear, and other far wider limits that a cold-blooded animal may endure and yet live. The effect of too extreme a cold is to lessen metabolism, and hence to lessen the production of heat. Both catabolic and anabolic changes share in the depression, and though less energy is used up, still less energy is generated. This diminished metabolism tells first on the central nervous system, especially the brain and those parts concerned in consciousness.

Both heart-beat and respiration-number become diminished, drowsiness supervenes, becoming steadily deeper until it passes into the sleep of death. Occasionally, however, convulsions may set in towards the end, and a death somewhat similar to that of asphyxia takes place. In some experiments on cats performed by Sutherland Simpson and Percy T. Herring, they found them unable to survive when the rectal temperature was reduced below 16°C. At this low temperature respiration became increasingly feeble, the heart-impulse usually continued after respiration had ceased, the beats becoming very irregular, apparently ceasing, then beginning again. Death appeared to be mainly due to asphyxia, and the only certain sign that it had taken place was the loss of knee jerks.

On the other hand, too high a temperature hurries on the metabolism of the various tissues at such a rate that their capital is soon exhausted. Blood that is too warm produces dyspnoea and soon exhausts the metabolic capital of the respiratory centre. The rate of the heart is quickened, the beats then become irregular and finally cease. The central nervous system is also profoundly affected, consciousness may be lost, and the patient falls into a comatose condition, or delirium and convulsions may set in. All these changes can be watched in any patient suffering from an acute fever. The lower limit of temperature that man can endure depends on many things, but no one can survive a temperature of 45 °C (113 °F) or above for very long. Mammalian muscle becomes rigid with heat rigour at about 50° C., and obviously should this temperature be reached the sudden rigidity of the whole body would render life impossible. H.M. Vernon has done work on the death temperature and paralysis temperature (temperature of heat rigour) of various animals.

He found that animals of the same class of the animal kingdom showed very similar temperature values, those from the Amphibia examined being 38.5 °C, Fishes 39 °C, Reptilia 45 °C, and various Molluscs 46 °C. Also in the case of Pelagic animals he showed a relation between death temperature and the quantity of solid constituents of the body, Cestus having lowest death temperature and least amount of solids in its body.

But in the higher animals his experiments tend to show that there is greater variation in both the chemical and physical characters of the protoplasm, and hence greater variation in the extreme temperature compatible with life.

Control of Temperature in Homeotherms

Animals capable of temperature regulation within a given range are deemed homeotherms (alternatively homiotherms or homotherms). They have the ability to regulate temperature via negative feedback control. Temperature is controlled in a variety of ways in these animals. The hypothalamus once again acts as a receptor in regulation, by detecting fluctuations in temperature.

These receptors are better known as thermoreceptors. Skin also possesses thermoreceptors which can detect the temperature of the external environment. This information is relayed to the hypothalamus which can in turn transmit nerve pulses for corrective mechanisms to occur

CORRECTIVE MECHANISMS IN TEMPERATURE CONTROL

Increased sweating is a corrective response aimed to reduce the temperature of the organism. Vasodilation is a corrective response where the blood vessels close to the skin surface become more dilated, meaning their is a larger surface area for heat to be lost of the external environment from the blood vessel carrying over-heated blood. Vasoconstriction is the opposite of this and occurs when temperatures in an organism drop.

The blood vessels become constricted so that minimal heat loss occurs. The hairs on your body also play an important role in temperature regulation. A corrective response can occur where the hairs 'stand on end', and trap a layer of air between the hair and the skin. This insulation of warmer air next to the skin reduces heat lost, while a thin layer of insulation would increase heat loss.

Other corrective mechanisms are involved, such as a drop in metabolic rate and shivering when temperatures drop. The pages have described the importance for regulation and equilibrium internal to organisms.

UNDERSTANDING PROTEIN FRACTIONS

Dietary crude protein (N% x 6.25) is not a very useful measure of what potentially happens with consumed nitrogen relative to utilization by the rumen or cow. Crude protein can be chemically separated into fractions on the basis of rumen degradability and solubility and have nutritional relevance to meeting the protein needs of the rumen and cow. Proteins that are rumen degradable would be able to provide nitrogen for microbial protein production. Rumen solubility suggests that the protein source would be more rapidly available. For example, urea, a non-protein nitrogen source, is 100% soluble and degradable and therefore would very rapidly provide ammonia for microbial protein production. Rumen insoluble and very slowly degraded is that dietary protein that can bypass to the abomasum to be digested enzymatically by the cow. The amount of unavailable dietary protein is measured as the amount of nitrogen found in the ADF fraction.

In meeting our goal of maximum microbial protein production, we need to match carbohydrate and protein rates of degradation. This allows for somewhat equivalent amounts of rumen energy and nitrogen availability promoting efficient microbial protein yield and overall dietary protein incorporation.

If we can reduce the amount of protein used in the ration and yet maintain or improve milk yield, the dairy cow becomes much more efficient, profitable and environmentally friendly! Protein Terminology - the following is a summary list of commonly used terms relative to protein nutrition of ruminant animals.

- *Crude Protein (CP)*: Total nitrogen per cent of a feed times 6.25 factor. Crude protein analysis does not differentiate between true protein and non-protein nitrogen sources.
- *Digestible Protein (DP)*: Amount of protein absorbed from the intestine from a feed source. Digested protein in the ruminant animal comes from microbial protein and undegraded dietary protein.
- *Soluble Protein (Sol CP, SIP)*: Dietary protein that readily goes into solution in rumen fluid. This is a rapidly available source of NH_3 for the rumen bugs. Composed of NPN and soluble true proteins.
- *Rumen Degradable Protein (RDP, DIP)*: Dietary protein that can be fermented in the rumen by microbes and contribute to the rumen ammonia pool. Rate of passage of digesta through the rumen will determine the extent of which RDP will be degraded.
- *Rumen Undegradable Protein (RUP, UIP), Escape Protein, Bypass Protein:* Protein which is either very slowly degraded in the rumen or which is unavailable to rumen fermentation. This protein is available for digestion in the abomasum to provide amino acids for the cow.
- *Metabolizable Protein (MP)*: Dietary protein (feed and microbial) that has been digested and is available for absorption and utilization.
- *Non-protein Nitrogen (NPN):* Nitrogenous compounds that do not contain linkages of amino acids. Potential protein source for rumen microbes but the utilization is dependent upon dietary CHO availability. Sources include urea, ammonia, biuret and amino acids.

- *Unavailable Protein, Bound Protein (ADIN, ADF-N)*: Protein that is unavailable to rumen degradation or abomasum digestion. This is the amount of nitrogen present in the ADF residue and is a measure of protein bound to fibre due to heat damage. It is very indigestible and reduces feedstuff quality.

Intestinal Protein Digestion

Methods for Measuring Intestinal Protein Digestion The total amount of protein available for absorption from the small intestine depends on the flow of microbial and dietary protein to the duodenum and their respective intestinal digestibilities.

Digestion of protein that leaves the rumen starts in the abomasum with acid-pepsin digestion and is completed in the small intestine with pancreatic and intestinal proteases. The NRC (1989) recognized that intestinal digestion of protein supplements may differ; however, empirical data were lacking, and as a result, a constant value of 80% was used for all feeds. The main problem was the lack of reliable techniques for estimating intestinal digestion of proteins (Stern *et al.*, 1997). *In vivo* estimation of intestinal protein digestion involves expensive and labor-intensive experiments and requires the use of surgically prepared animals. Apparent digestion of protein is calculated as the disappearance of CP or amino acids between the duodenum and ileum, which is subject to considerable error associated with digesta sampling, use of digesta flow rate markers, and inherent animal variation. Therefore, several alternative procedures for estimation of intestinal protein digestion in ruminants have been developed.

Making a Dynamic Rumen Model

Over the past few decades much research has been devoted to quantification of total dietary nutrient needs of dairy cows to support all levels of performance. Many computerized feeding programmes a based on comparison between nutrients delivered in the diet and defined daily nutrient requirements. However, this approach is not sensitive to dynamic changes in feed intake and other factors that impact feed degradability and availability, which ultimately impact the amount of metabolizable nutrients available to the cow to support production.

Nutrient availability from a feed either by microbial fermentation or intestinal digestion is not constant. Competition between inherent feed properties of degradability or digestibility and rate of passage through the digestive tract can explain the amount of a given nutrient pool will be available.

This interaction of rates is mathematically described in equations. This concept is the central component of any dynamic rumen model such as the Cornell Net Carbohydrate and Protein System (CNCPS).

(1) Amount Degraded = Pool size x K_d / [K_d + K_p]

(2) Amount Passed = Pool size x K_p / [K_d + K_p]

How much of a given carbohydrate or protein fraction within a feed is degraded will depend upon how rapid the rate of degradation is compared to rate of passage. Rate of passage through the rumen depends upon the fraction. Forages or fibre are slowly passed between 4 and 8 per cent/hr whereas the liquid fraction passes more quickly, between 12 and 20 per cent/hr. If a feed fraction has a rate of degradation that greatly exceeds rate of passage, then most or all of the fraction will be degraded in the rumen. If rate of degradation is slower than rate of passage, then only a small fraction will be degraded in the rumen.

The dynamic components to this system are those CHO or protein fractions where their rate of degradation is nearly equal to the rate of passage. For these fractions, *i.e.*, fermentable fibre CHO and insoluble degradable protein, slight changes in either rate of passage or degradation can greatly influence how much of the fraction will be degraded in the rumen. The following describes a variety of factors that can influence ruminal degradation and passage rates.

Rate of Degradation (K_d)

Degradation of a feed ingredient is inherently related to the chemical and physical properties of all component compounds. Relative to CHO feeds, degradation (ruminal or intestinal) of sugars is extremely rapid compared to starch or fibre. Within starches there is some range of degradation. Fibre is more slowly fermented, while some fibre is not fermented at all. Similarly with proteins, soluble proteins are rapidly degraded while others are more slowly or not at all. It is important then to have an accurate chemical analysis of feed ingredients to determine how much of each of these different fractions comprise a given feed ingredient.

Beyond chemical and physical properties of a feed, degradability can be modified by feed processing, grinding, and rumen conditions. Heating, grinding, ensiling or some combination will increase rumen degradability of feed CHO starch and fibre fractions. For example, corn starch is very crystalline in nature and hydrophobic, thus being fairly resistant to fermentation without processing. If corn grain is ensiled and ground starch availability will be increased dramatically.

This explains why coarsely ground corn may be seen in manure compared to ground high moisture corn causing acidosis. Ensiling will increase protein availability to the rumen through increasing solubility. Heat treatment of protein makes the insoluble degradable fraction less rumen degradable. Proteolytic activity in the rumen will dictate extent of rumen protein breakdown of degradable fraction. A decline in rumen pH will greatly reduce fibre fermentation and may reduce proteolytic activity.

Rate of Passage (K_p)

Rate of ruminal passage is calculated for forages and concentrates. Rate of passage will always be slower for forages compared to concentrates. Dry matter intake, forage concentration in the diet and body weight are the primary determinants of rate of passage. Rate of passage increases with increasing dry matter intake and dietary forage concentration. Rate of passage declines slightly with increasing body weight. For each feed an adjustment factor to rate of passage is calculated based on effective NDF of the feed. Passage rate will be reduced with higher eNDF content of feeds.

Putting the Concepts Together

When designing a ruminant feeding programme, concepts of rates need to be addressed in attempting to make sure ruminal availability of energy and nitrogen are coordinated in order to achieve maximal microbial protein production. The goal is to achieve some level of synchrony between ruminal energy and nitrogen sources in order to maximize microbial protein yield. If energy and nitrogen sources are not synchronized, then loss of available energy or nitrogen and reduced microbial activity will occur. Asynchronous energy and nitrogen sources will result in reduced milk production and efficiency and elevated blood urea nitrogen values. In this scenario diseases such as urea toxicity, grain overload or other rumen dysfunctions may occur. If energy and nitrogen sources are synchronized, we maintain maximal microbial protein synthesis. In this scenario not only will the amount of additional dietary protein be reduced, but the cows will have increased dry matter intake and remain healthier, all contributing to increased milk producing efficiency. How is this synchronization achieved?

Intuitively, synchronization of ruminal energy and nitrogen sources is achieved by utilizing various CHO and protein feed ingredients that have differing degradability properties. Use of a single protein (canola meal or soybean meal) or energy source (corn grain) does not give one this flexibility. Using ensiled feeds as the sole forage programme unbalances protein to the more readily degradable fractions. One can alter degrability properties with grinding, heating or some other processing procedure as previously discussed. This is where the strength of the dynamic formulation programmes comes into place. One can use such software to appropriately balance feed ingredients to best meet ruminal needs. In addition, one can use milk or blood urea nitrogen as a diagnostic aid in evaluating the balanced achieved between dietary CHO and nitrogen sources.

Another more simplified way of achieving relative synchrony to the rumen environment is through the feeding programme. Feeding a total mixed ration (TMR) is the single best method of approximating rumen nutrient synchrony. The needed balance between CHO and protein fractions is achieved with multiple meal feeding. This effect accounts for the observed 5-8% improvement in milk production when shifting a conventional to a TMR feeding programme. More slowly degraded CHO fractions can trap readily available nitrogen at a later meal. When concentrates are fed in the milking parlor, this balance is lost. In a conventional feeding system where feed ingredients are fed separately, one then needs to more closely account for differences in CHO and protein availability within feeds.

Thus, sequence of feeding can have an impact on milk production response. Forages should be fed prior to concentrate meals to maintain slower rate of passage. Large concentrate meals are best divided and fed 3 to 4 times or more per day to limit the effect of acidosis.

FEEDING RECOMMENDATIONS FOR DIETARY CARBOHYDRATES AND PROTEIN

The preceding discussion has focused on understanding the rumen system and presented concepts on how to better feed the microbial populations to improve milk production efficiency. To move into this new era of dairy cattle feeding, more comprehensive methods of feed composition analysis are needed.

With a better understanding of feed composition, we can apply concepts of rates of degradation and passage to a dynamic rumen system to achieve improved feeding practices that will not only improve milk production efficiency, but decrease wasted dietary nitrogen from entering the environment. The following tables provide some practical feeding guidelines in applying these concepts to ration formulation.

NDF and Dry Matter Intake

As previously described, NDF has been shown to influence dry matter intake in ruminant animals. Mertens had shown optimum intake of NDF as a per cent of body weight (%BW) was 1.2 " 0.1 %BW. Neutral detergent fibre intake capacity is influenced by physiologic state and age. Cows can consume more or less dietary NDF, however, there will be negative consequences on production or performance as a result of reduced intake.

Non-fibre Carbohydrates

The NFC portion of the diet is primarily responsible for rapid ruminal fermentation in support of milk production. Milk production will be severely compromised in the face of low dietary NFC content. Conversely, excess dietary NFC content may result in ruminal dysfunction, namely subclinical or clinical acidosis.

The balancing act with NFC is to provide sufficient amounts to support maximal microbial growth, yet not providing excess amounts that will reduce ruminal pH and compromise fermentation activity. Table presents suggested dietary concentrations for total NFC and NSC (sugar and starch) components over differing physiologic states.

Protein Fractions

Table. Suggested Dietary Concentrations for Various Protein Fractions in Diets for Dairy Cattle.

	Crude Protein	Soluble CP	RDP[1]	RUP[1]
Dry Cows	% DM	% of CP	% of CP	% of CP
Far Off	12 - 13	35	70 - 72	28 - 30
Close-up	14 - 15	30	63 - 65	35 - 37
Lactating Cows				
Fresh	19	30	60 - 62	38 - 40
22-80 DIM	18	31	64	36
81-200 DIM	16	32	66	34
>200 DIM	14	34	68	32

Abbreviations: RDP = rumen degradable protein; RUP = rumen undegradable protein. In formulating for dietary protein, rumen nitrogen needs must be considered first. Ruminal degrable protein should be between 10.5 and 11% of dry matter to support microbial nitrogen needs. One can never separate rumen nitrogen from CHO needs. To address this interaction between dietary CHO and protein in the rumen, one should formulate for a NFC:RDP ratio between 3.0 - 3.5 to 1. This suggests as NFC content goes up, then more degradable protein can be assimilated.

From a dietary protein formulation perspective, nitrogen needs of the rumen should be met first, then metabolizable protein requirements for the cow should be addressed. Dietary crude protein is a poor method by which to formulate diets in this manner. However, one needs some sophisticated dynamic models to adequately predict dietary metabolizable protein delivery.

If one formulates diets to the specific protein fractions, the amount of metabolizable protein should be reasonably adequate. Again, one can only truly predict the degradability and undegradablility of protein sources with a dynamic modeling system, which accounts for dry matter intake and ingredient interactions on availability and contribution to either microbial protein yield or digestible undegraded protein.

Rumen protein needs should be met with forage and plant source proteins. If additional bypass protein is necessary, then heat-treated plant proteins or animal proteins (if allowable) can be used to meet the cow's metabolizable protein needs. Use of rumen protected amino acids should be left to only fine tune a diet if necessary, given their cost.

5

Respiratory in Animals

All animals need to take in O_2 and eliminate CO_2. *Lungs* are membranous structures designed for gas exchange in a terrestrial environment. *Gills* are designed for gas exchange in an aquatic environment. Oxygen must be dissolved in water before animals can take it up. Therefore, the respiratory surfaces of animals must always be moist. This is true of all animals. Very small organisms donot need respiratory surfaces because they have a high surface:volume ratio.

Skin

The skin can be used as a respiratory surface but it does not have much surface area compared to lungs or gills. Animals that rely on their skin as a respiratory organ are small and either have low metabolic rates or they also have lungs or gills. Like all respiratory surfaces, the skin must remain moist to function in gas exchange. Amphibians, most annelids, some mollusks, and some arthropods use their skin as a respiratory organ.

Gills

Gills provide a large surface area for gas exchange in aquatic organisms. It is difficult to circulate water past gills because water is dense and the O_2 concentration in water is low. There is 5% as much oxygen in water as there is in air. To circulate water past the gills, amphibian larvae physically move their gills, mollusks pump water into mantle cavity which contains the gills, and some crustacean gills are attached to branches of the walking legs. The flow of blood in the gills of fish is in the opposite direction that water passes over the gills. This arrangement enables fish to extract more oxygen from the water than if blood moved in the same direction as the passing water. Gills cannot be used in air because they lack structural support; they would collapse. Their use in air would also result in too much water loss by evapouration.

Gills greatly increase the surface area for gas exchange. They occur in a variety of animal groups including arthropods, annelids, fish, and amphibians. Gills typically are convoluted outgrowths containing blood vessels covered by a thin epithelial layer. Typically gills are organized into a series of plates and may be internal or external to the body. Gills are very efficient at removing oxygen from water: there is only 1/20 the amount of oxygen present in water as in the same volume of air. Water flows over gills in one direction while blood flows in the opposite direction through gill capillaries. This countercurrent flow maximizes oxygen transfer.

Tracheal System

Insects, centipedes, and some mites and spiders have a tracheal respiratory system. Tracheae are a network of tubules that bring oxygen directly to the tissues and allow carbon dioxide to escape. The openings to the outside, called spiracles, are located on the side of the abdomen. Trachea and lungs are internal to reduce water loss. Many terrestrial animals have their respiratory surfaces inside the body and connected to the outside by a series of tubes.Tracheae are these tubes that carry air directly to cells for gas exchange. Spiracles are openings at the body surface that lead to tracheae that branch into smaller tubes known as tracheoles. Body movements or contractions speed up the rate of diffusion of gases from tracheae into body cells. However, tracheae will not function well in animals whose body is longer than 5 cm.

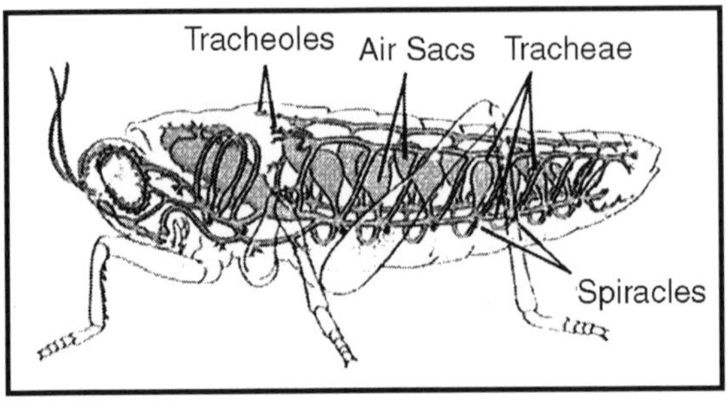

Fig. ***Respiratory System in an Insect.***

Vertebrate Lungs

- Simple lungs evolved 450 million years ago in fish.
- Some evolved into swim bladders.
- Others evolved into more complex lungs.
- Paired lungs are the respiratory surfaces in all reptiles, birds, and mammals.

Amphibians

Lung is a simple convoluted sac have small lungs but obtain much O_2 by diffusion across moist skin ventilate lungs by positive pressure; (reptiles, birds and mammals use negative pressure)

Reptiles

- The skin is watertight; it is not used as a respiratory surface.
- The lungs possess alveoli.
- All diffusion occurs across the alveolar surface.

Birds and Mammals

The lungs of birds and mammals are more branched with smaller, more numerous alveoli.

Birds

Birds have one-way flow of air in their lungs. As a result, the lungs receive fresh air during inhalation and again during exhalation. *Advantages of one-way flow*: no residual volume; all old (stale) air leaves with each breath crosscurrent flow (crosscurrent = 90° ; countercurrent = 180°; crosscurrent is not as efficient but is still more efficient than mammalian lung) One-way flow is accomplished by the use of air sacs as illustrated below. During inspiration, the air sacs fill. During expiration, they empty.

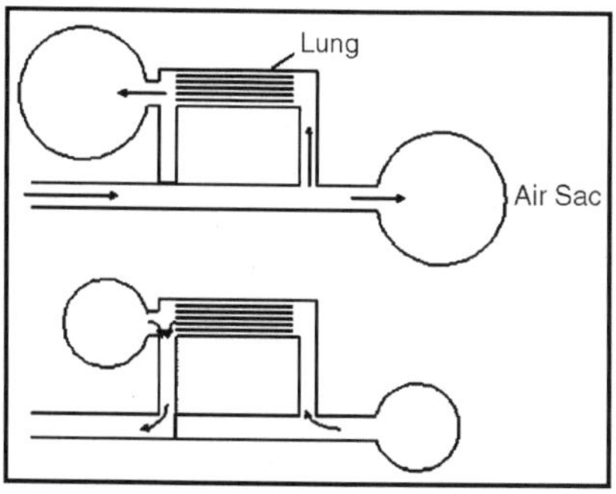

RESPIRATORY SYSTEM

Oxygen is needed by aerobic organisms because it is the final electron acceptor during cellular respiration. The diagram below shows that cellular respiration is a process in which electrons are removed from glucose in a series of steps. The electrons are carried by NADH and $FADH_2$ to the electron transport system. The electron transport system uses the energy in the electrons to synthesize ATP. The remaining carbon atoms in the glucose molecule are released as CO_2, a waste product. The equation for the complete breakdown of glucose by aerobic eukaryotes is:

$$C_6H_{12}O_6 + 6O_2 = 6CO_2 + 6H_2O + 36\ ATP$$

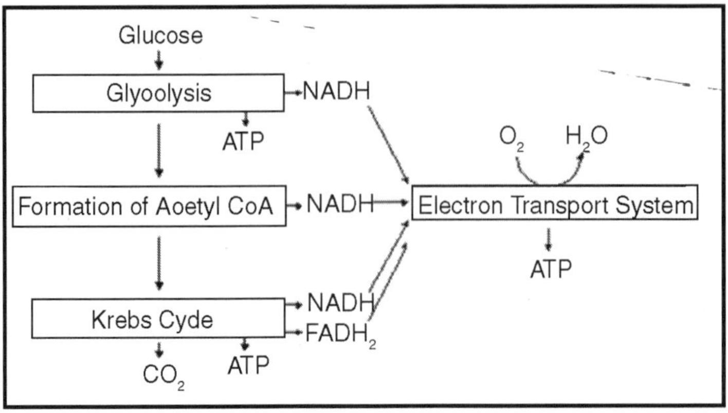

78% N_2, 21% O_2, 1% argon, noble gases, CO_2

Diffusion refers to movement of molecules from an area of higher concentration to an area of lower concentration. Partial pressure is the pressure exerted by one gas in a mixture. Total atmospheric pressure at sea level = 760 mm Hg. Partial pressure $O_2 = 760 \times .21 = 160$ mm Hg. Gasses move by diffusion from areas of higher partial pressure to areas of lower partial pressure.

The respiratory system

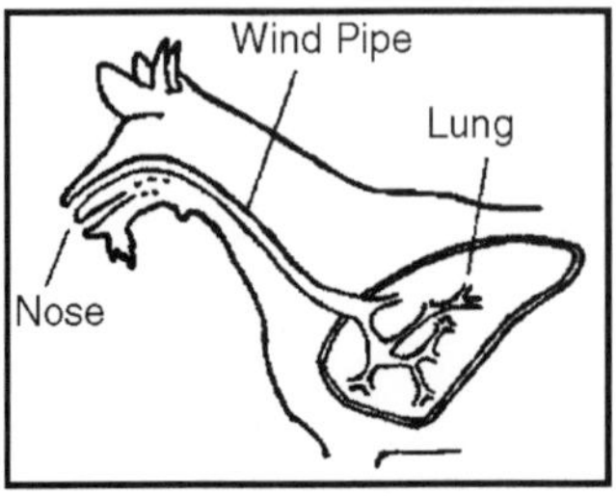

Flatworms and annelids use their outer surfaces as gas exchange surfaces. Earthworms have a series of thin-walled blood vessels known as capillaries. Gas exchange occurs at capillaries located throughout the body as well as those in the respiratory surface.

Amphibians use their skin as a respiratory surface. Frogs eliminate carbon dioxide 2.5 times as fast through their skin as they do through their lungs.

Eels (a fish) obtain 60% of their oxygen through their skin. Humans exchange only 1% of their carbon dioxide through their skin. Constraints of water loss dictate that terrestrial animals must develop more efficient lungs.

Lungs

Lungs are ingrowths of the body wall and connect to the outside by as series of tubes and small openings. Lung breathing probably evolved about 400 million years ago. Lungs are not entirely the sole property of vertebrates, some terrestrial snails have a gas exchange structures similar to those in frogs.

Lungs and Pleura

Palpation - The only evidence of disease manifested on palpation is that of pain. Percussion - Sound on percussion is best noted over thin chest walls. The sound is resonant normally and is clearer about the middle of the chest in large animals.

The sound is clear through-out in small animals. In some diseases, the resonance is more extensive and very clear such as Pulmonary Emphysema or Pneumothorax.

The sound is dull and flat over area of consolidation in Pneumonia, in tumors or cysts, in Hydrothorax or in Pleurisy with effusion. In the latter two conditions, the dull sound is made outing the lower portion of the chest up to the upper level of the fluid. The outline of the dull areas is irregular in Pneumonia.

Auscultation

The sounds of lungs are intensified by exercise. The normal sounds are:

1. Vesicular murmur
2. Bronchial sounds

The vesicular murmur is a short sound noted specially on inspiration and is due to stretching of the walls of the vesicles. The vesicular murmur is intensified in dyspnoea and in bronchitis in' the early stage. It is diminished in thickening of thoracic walls and in pulmonary congestion.

It is absent over areas of consolidation in pneumonia, tumours or cysts and over pleural exudates ortransudates. In the latter case; the absence of sound is due partly to displacement of the lungs and partly to collapse. In pneumonia, the solid area is surrounded by a zone in which feeble crepitating sound may be heard.

The bronchial sounds are formed by the larynx. They are blowing sounds and most marked at the middle portion of the lungs. The bronchial sounds like the other lung sounds are synchrony-nous with respiration and most marked during inspiration. Various abnormal bronchial sounds are heard and are called 'Bales' and distinguished as dry and moist.

Dry Rales

Are due to passage of air over tough bronchial secretion or over swollen bronchial mucous membrane. They are humming, hiss-in, wheezing or whistling and are noted in the early stages of Acute Bronchitis, Chronic Bronchities and, to a slight extent, in Emphysema or in Compression of the bronchi with nodules or cysts.

Moist Rales

These rile are gurgling or bubbling. They appear in the stage of exudation in bronchitis and are due to the passage of air through or over the fluid present. They are most marked towards the end of inspiration.

The Crepitate Rales

Fine crackling noises due to separating of adhesive walls of the small bronchial tubes. They are noted in bronchiolitis, in pulmonary edema and in the early stage and the stage of resolution in pneumonia. In the latter case, they are often called secondary crepitationsounds. In Acute Pleurisy, during the early stages, that is, in the dry stages, friction sounds may behead comparable with the rustling of silk due to rubbing of dry pulmonary pleura on the costal pleura. These sounds are synchronous with the respiration. When the fluid accumulates in the pleural sac, trickling or splashing sounds synchronous with the respiration at the upper level of the fluid may be noted. In examination of the lungs, Borborygmous and the sounds produced by regurgitation of food or gas through the esophagus must be distinguished.

Respiratory System Principles

- Movement of an oxygen-containing medium so it contacts a moist membrane overlying blood vessels.
- Diffusion of oxygen from the medium into the blood.
- Transport of oxygen to the tissues and cells of the body.
- Diffusion of oxygen from the blood into cells.
- Carbon dioxide follows a reverse path.

URINARY SYSTEM

The main organs are the two kidneys, which lie against the backbone, and the bladder. Waste materials and water are taken out of the blood in the kidneys. This forms urine. Urine collects in the bladder then passes out of the body.

Primary functions of the urinary system include:

- Excretion of waste products of metabolism;
- Maintenance of a constant extracellular environment through conservation and excretion of water and electrolytes;
- Production of the hormones erythropoietin and renin, which regulate hematopoiesis, blood pressure, and sodium reabsorption;
- Metabolism of vitamin D to its active form (1,25-dihydroxycholecalciferol).

Many abnormalities of the urinary system can be diagnosed from the signalment of the patient, history and physical examination findings, serum chemistry profile, urinalysis, and aerobic bacterial urine culture. The history should include information regarding changes in water consumption, frequency of urination, volume of urine produced, appearance of urine, and behaviour of the patient. It is also important to obtain information about historical and current drug administration, appetite, diet, changes in body weight, and previous illnesses or injuries. The physical examination should include palpation of the bladder and examination of external genitalia.

In dogs, rectal examination should be performed to evaluate the urethra in both sexes and for evaluation of the prostate in male dogs. Rectal examination in cats may not be feasible due to their small size; however, the kidneys are generally easier to palpate in cats than in dogs. A full neurologic examination should be performed on all animals with micturition disorders. Additional diagnostic tests, such as CBC, blood gas analysis for acid-base status, blood pressure, urine protein:creatinine ratio, iohexol clearance test, survey abdominal radiography, abdominal ultrasonography, contrast studies of the upper and lower urinary tract, cystoscopic examination of the urinary bladder, and renal biopsy may also provide valuable information.

Protein in urine should be evaluated in light of the urine specific gravity. Protein in a concentrated urine sample may not be significant, whereas the same amount in a dilute sample may be significant. Urine dipsticks provide a semiquantitative assessment of protein and can be influenced by urine pH. Therefore, they should be used only as a screening test for protein, not as a definitive diagnosis of proteinuria. A urine protein:creatinine ratio from a single urine sample or from a 24-hr urine sample is required to quantitate the amount of protein in urine. In dogs, the following guidelines should be used for interpretation of urine protein:creatinine ratios: 0.0-0.3 = normal; 0.3-1.0 =questionable; and>1.0 = abnormal. In cats, a urine protein:creatinine ratio<0.7 is considered normal. Urine protein:creatinine ratios must be interpreted in the context of other information from the urinalysis. Inflammation and hematuria can falsely elevate urine protein:creatinine ratios. The urinary system is responsible for filtering wastes from the blood and both forming and secreting urine. These functions help to maintain the composition and volume of body fluids.

Although it has far-reaching effects, the urinary system is relatively simple anatomically and consists of:

- Kidneys
- Ureters
- Bladder
- Urethra

The main organs are the kidneys, which filter blood and produce urine. The other parts are simply accessory structures for the transport and storage of urine. During the normal breakdown of protein and nucleic acids, nitrogen is released into the bloodstream. Some of this nitrogen is recycled to make new cellular products, but most of it is disposed of. The body has to have a way to rid itself of this unused nitrogen, as high levels in the blood can be toxic.

Most of the nitrogen is bound with hydrogen as NH_3 (ammonia), which is readily dissolved in water. For this reason, fish are able to excrete much of their nitrogen by simple diffusion into the surrounding water. The build-up of nitrogen in the water is one of the reasons that tank water needs to be changed regularly. Terrestrial animals have a different way of ridding their bodies of excess nitrogen. It is either excreted as uric acid or urea. Animals that are concerned about water loss, such as birds and reptiles, excrete the more concentrated uric acid as a pasty white material. Mammals, on the other hand, can excrete urea, along with more water. The mixture of urea, water, and other wastes is called 'urine.' Urine is still very concentrated in comparison to the blood, and the system that facilitates this concentration is the 'urinary system'.

A urinalysis is unreliable for ruling out a urinary tract infection (UTI). Not all UTI are associated with an inflammatory response. In addition, >10,000 bacterial rods/mL and >100,000 bacterial cocci/mL of urine are required to consistently find bacteria in a urine sample using light microscopy. Urine samples for bacterial culture may be obtained by the same methods used for obtaining samples for urinalysis; however, the preferred method is cystocentesis. Urine obtained by cystocentesis should be sterile. If urine samples are collected by methods other than cystocentesis, a quantitative urine culture should be requested. If the sample is collected by spontaneous micturition or manual compression, significant numbers of bacteria are present if e"100,000 colony forming units (CFU)/mL of urine in dogs or e"10,000 CFU/mL of urine in cats are detected. Samples with >10,000-90,000 CFU/mL in dogs and >1,000-10,000 CFU/mL in cats are suspicious for a UTI. If the sample is collected by catheterization, e"10,000 CFU/mL in dogs and e"1,000 CFU/mL in cats is significant, while samples containing 1,000-10,000 CFU/mL in dogs and 100-1,000 CFU/mL in cats are suspicious for a UTI.

More sensitive methods for detecting renal dysfunction include plasma clearance tests (*e.g*, insulin clearance), radionuclide techniques, endogenous creatinine clearance, and exogenous creatinine clearance. However, these tests are impractical to perform routinely in clinical practice. The iohexol clearance test is a recently developed alternative for detecting renal dysfunction. It entails recording an accurate body weight, administering a precise amount of iohexol IV, and accurately timing collection of blood samples as directed following administration. This test does not require timed collection of urine samples or special equipment.

Depending on the cause of the urinary tract disorder, radiographic procedures, sonographic examination, and cystoscopic examination of the bladder may provide additional valuable information. The kidneys have a limited range of responses to disease; therefore, renal biopsies are rarely useful when evaluating renal dysfunction. An exception to this is in animals with significant proteinuria. Blood gas analysis or serum bicarbonate levels provide useful information on acid-base status, especially in animals with renal dysfunction. Metabolic acidosis is a common problem in chronic renal failure and can result in protein catabolism.

Kidneys

The kidneys of mammals are round or bean-shaped organs. They are located outside of the peritoneum – the membrane that encloses the organs of the abdominal cavity. Because of this position, they are referred to as 'retroperitoneal'. They are surrounded by fat tissue known as 'perirenal fat'. A fibrous capsule covers the kidney. The indentation of the bean shape is called the 'hilum.' The hilum is the site where the renal artery enters the kidney and both the renal vein and ureter exit.

The kidney can be divided into two distinct regions – the outer cortex and the inner medulla. The cortex is where blood is actually filtered through small structures called 'glomeruli'. The medulla is where the urine is concentrated through a complex system of tubules. They accomplish this by absorbing the water and electrolytes while preventing waste products from being reabsorbed. One glomerulus and its corresponding set of tubules are called a 'nephron' – the microscopic functional unit of the kidney.

The tubules of multiple nephrons are grouped into larger visible portions of the kidney called 'pyramids'. The renal columns are spaces between the renal pyramids that provide a route for blood vessels traveling to the cortex. The tips of the pyramids are called 'papillae,' and they drain urine from nephron tubules into larger vessels called 'minor calyxes'. The minor calyxes converge into still larger vessels called the 'major calyxes'. These lead to the enlarged opening of the ureter. This collecting chamber is called the 'renal pelvis.'

Nephrons

The nephrons are the actual filtration elements of the kidneys. Blood enters the kidneys through the renal artery and travels through branching arteries. These arteries coming into the nephron become smaller and smaller, and are finally called 'afferent arterioles.'

The blood in these vessels is under high pressure. The afferent arteriole branches into the glomerulus - a cluster of capillaries within a shell called 'Bowman's capsule.' The pressure of the blood causes water, glucose, amino acids, and salts to leave the blood vessels and enter the Bowman's capsule. The blood cells and most proteins are too big to be filtered out, and remain inside the blood vessel to be carried out of the glomerulus via the efferent arteriole.

Bowman's capsule leads to the network of tubules that concentrate the filtrate into urine. The tubules are surrounded by small blood vessel called a 'capillary'. This capillary is where materials are reabsorbed back into the blood. The first part of the tubule is aptly named the 'proximal convoluted tubule' for its twisted shape. Here 99% of the water is reabsorbed along with all of the glucose and amino acids. The presence of glucose or amino acids in the urine is a sign of disease. For example, diabetics that have too much glucose in their blood cannot reabsorb it all, so it is excreted in the urine.

The proximal convoluted tubule leads to the 'loop of henle,' a long looping structure that extends down into the medulla of the kidney. More water and electrolytes (salts) are absorbed here. Next, the filtrate is passed through the 'distal convoluted tubule,' where excess potassium ions, hydrogen ions, and some drugs or toxins are passed from the blood into the filtrate. The final product is then dumped into a large collecting duct, into which several nephrons empty. This collecting duct leads to the papillae of the pyramids, through the calyxes, and into the renal pelvis to be excreted through the ureter.

Ureters

The ureters are muscular tubes that transport urine from the kidneys to the urinary bladder.

Ureters have three layers of tissue:

1. Fibrous outer coat
2. Muscular layer
3. Inner mucosal layer

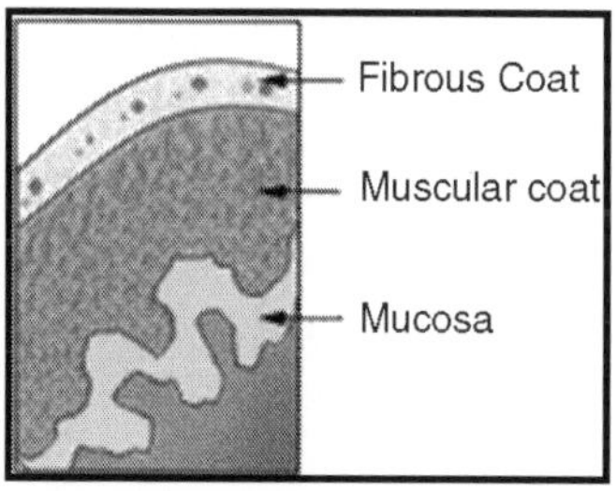

The muscle layer is the functional layer, using peristalsis to move the urine along. Peristalsis is a waving contraction of the muscles to propel the contents of a tube in one direction. In this case, the urine is propelled to an opening at the base of the bladder.

Urinary Bladder

The urinary bladder is a sac for the temporary storage of urine. It is located in the pelvic cavity. The outer surface is covered with fibrous connective tissue. Inside the connective tissue is a muscular layer called the 'detrusor muscle.' This smooth muscle contracts to expel urine from the bladder.

The next tissue layer is the 'submucosa' an elastic fibrous membrane that supports the mucosa, which lines the inside of the bladder. The mucosa is composed of specialized cells called 'transitional epithelium.' When the bladder is empty, the mucosa has many folds termed 'rugae.' The rugae and transitional epithelium allow the bladder to stretch when filled with urine.

At the base of the bladder, a triangular structure called the 'trigone' is formed from the openings of the two ureters and the urethra. The opening of the urethra is surrounded by a band of detrusor muscle, forming an internal urethral sphincter. This sphincter is relaxed by involuntary muscle control, and is innervated such that when the bladder is approximately half full, the animal perceives the urge to urinate.

Urethra

The final passageway for urine out of the bladder is through the thin-walled urethra. This tube runs from the base of the urinary bladder to the outside of the body. In the female, it is relatively short, connecting the bladder to external urethral sphincter. In males, however, it is longer.

It passes through the prostate gland (in those animals that have one) and travels the length of the penis before reaching the external sphincter. The external urethral sphincter is voluntarily controlled, and is relaxed by the animal when a suitable site and time for urination has been determined.

PROCESS OF RESPIRATORYSYSTEM

Respiration (breathing) consists of inspiration (breathing in) and expiration (breathing out). There are two lungs which are found in the chest protected by the bony cage of the ribs. The windpipe carries air from the nostrils to the lungs which are spongy because of air spaces in them. As the animal breathes, air moves in and out of the lungs. Inside the lungs oxygen needed by the body passes into the blood in the walls of the lungs and water and carbon dioxide pass out of the blood into the air which is then breathed out.

RESPIRATORY SYSTEM OF THE DOG

The respiratory system includes the:

- Mouth and nose
- Trachea
- lungs and smaller airways (bronchi and bronchioles)

The respiratory system is responsible for taking in oxygen and eliminating waste gases like carbon dioxide. Because dogs and cats do not sweat through the skin, the respiratory system also plays an important role in regulation of temperature.

TECHNIQUES OF DIAGNOSTIC

Clinical history and physical examination should help to determine the possible cause and site of respiratory disease. Lateral cervical and thoracic radiographs may be helpful when obstructive upper airway disease or fixed airway obstruction is suspected (eg, tracheal foreign body, masses, foreign bodies, or stenosis). Thoracic radiographs are essential in any animal exhibiting lower respiratory signs (eg, cough, rapid shallow breathing, dyspnea) but diagnostic value may be limited in animals that have a large thorax (eg, adult horse or cattle). Arterial blood gas analysis or pulse oximetry may help assess the need for oxygen therapy in an animal with severe dyspnea. When obstructive upper airway disease is suspected, the diagnostic procedure of choice is endoscopy of the respiratory tract, preferably without sedation. Laryngeal function should be assessed, and the presence of obstructive lesions within the nasopharynx, oropharynx, larynx, trachea, or principal bronchi should be identified.

With diffuse or lobar lung disease, diagnostic procedures include transtracheal wash, bronchoscopy with bronchoalveolar lavage or endobronchial biopsy, and transthoracic fine needle aspirates of lung or lung biopsy. When bacterial pneumonia is suspected, bacterial culture of transtracheal wash is recommended. Cytologic evaluation of transtracheal or bronchoalveolar lavage fluid may aid in the diagnosis of fungal, parasitic, or allergic lung diseases.

Transthoracic fine needle aspirates of lung often are useful in the diagnosis of fungal pneumonia but have lower yields in the definitive diagnosis of solitary pulmonary lesions. Solitary pulmonary masses often require transthoracic lung biopsy or surgical excision for definitive diagnosis. Transthoracic ultrasonography is a sensitive diagnostic tool for pleural disease (eg, pleural effusion, pneumothorax) and for parenchymal lung disease when lesions are adjacent to the pleural surface.

In dogs or cats with pleural effusions, thoracocentesis should be performed for cytologic and potentially microbiologic evaluation of fluid. In cats, pleural effusions often occur with cardiac disease, so an echocardiography should be performed. In animals suspected to have a chylous effusion, serum and fluid triglyceride levels should be determined. Chylous effusions are associated with fluid triglyceride levels greater than that in serum.

Acute nasal discharge, sneezing, or both, may suggest the presence of infection (viral or bacterial) or a nasal foreign body. Chronic nasal discharge warrants further investigation via radiography (nose, guttural pouches in horses), nasal CT, rhinoscopy, nasopharyngoscopy, or nasal biopsy. Rhinoscopy may be of limited value if copious thick discharge or hemorrhage is present.

Bacterial cultures of nasal tissue may be of value if bacterial rhinitis is suspected; however, in some species (eg, dogs and cats) primary bacterial rhinitis is rare and typically occurs secondary to other nasal conditions. Cytologic evaluation of nasal tissue may help diagnose nasal fungal infections. Serologic testing for fungal respiratory infections may be considered, but these findings should correlate with the animal's clinical signs and documentation of the presence of fungal organisms because false-positive and false-negative tests can occur.

RESPIRATORY MALFUNCTION

Congenital anomalies of the respiratory tract are rare but do occur. Examples include cysts in the sinuses and turbinates, tracheal hypoplasia, nasopharyngeal turbinates, and accessory lungs. A common cause of upper respiratory tract malfunction is rhinitis (which results in exudation of neutrophils, macrophages, and fluids), or erosion and ulceration (or both) of the nasal mucosa.

It may be caused by viral, bacterial, fungal, or parasitic agents, as well as by hypersensitivity reactions, such as localized allergies and anaphylaxis. Atrophy of the turbinates (eg, in atrophic rhinitis of pigs) removes a major filtration function and exposes the lungs to much heavier loads of dust and microorganisms. The nasal cavity may be obstructed by tumors, granulomas, abscesses, or foreign bodies. Sinusitis can be a complication of upper respiratory infections, tooth root infection, or dehorning.

Laryngitis, tracheitis, and bronchitis result in coughing and possibly inspiratory or expiratory dyspnea. Coughing may be non-productive if the irritation is caused by mucosal erosion, or productive if caused by copious exudate in the major airways. Severe pulmonary edema and emphysema cause extreme respiratory insufficiency.

The most common respiratory disease is pneumonia, which is defined as inflammation of the lungs. There are many systems to classify the various types of pneumonia. One useful method is to classify according to the distribution of lesions in the lungs. Focal pneumonia has one or more discrete foci in a random pattern, eg, abscessation due to emboli from other sites, tuberculosis, or actinomycosis.Lobular pneumonia accentuates the anatomic pattern of lobules, as in bronchopneumonia caused byPasteurella multocida. Lobar pneumonia covers large areas of lobes and is often severe (eg, fibrinous pneumonic pasteurellosis of cattle). Diffuse or interstitial pneumonia often involves the entire lung, as in maedi of sheep or in hypersensitivity reactions. The appearance or cause of a particular pneumonia can be described further, eg, gangrenous, parasitic (verminous), aspiration, etc.

Infection may develop as a result of one or a combination of factors: 1) defence mechanisms are overwhelmed, 2) the infectious agent is highly virulent, 3) the size of the inoculum is large, and/or 4) the animal's defence mechanisms are compromised. The initial problem in many pneumonias is thought to be a sudden change in the normal nasal bacterial flora, which results in a sudden dramatic increase in one or more species of bacteria. Bacterial proliferation is usually caused by a breakdown of the host defences as a result of stress (eg, transportation, concurrent illness) or cellular insult (eg, viral infection, toxicity). These bacteria are breathed into the lung in large numbers and may overwhelm the normal defence mechanisms, localize, multiply, and initiate inflammation. In addition, stress is often a precursor of viral respiratory infections, particularly in groups of animals that have recently been congregated and stressed by travel, handling, and mixing. Some respiratory viral infections can cause temporary dysfunction of phagocytic mechanisms of the alveolar macrophages. This usually occurs several days after viral exposure. Inhaled bacteria proliferate and pneumonia ensues, often with an overwhelming infection and massive exudation into the alveoli.

Pneumonia also can be caused by direct infection with viruses, bacteria, and fungi, as well as by toxins arriving hematogenously, by inhalation, or by aspiration of food or gastric contents.

Through natural processes, possibly aided by appropriate therapy, the exudate may be removed from the lungs, and the mucosal lesions of the air passages may heal. However, serious sequelae can persist.Bronchiectasis is a chronic lesion of the bronchi and parenchyma characterized by irreversible cylindrical or saccular dilatation, secondary infection, and atelectasis. Ulceration of bronchioles caused by viral agents may lead to organized plugs of connective tissue in small bronchioles, a lesion calledbronchiolitis obliterans, which may cause permanent obstruction, atelectasis, and severe respiratory insufficiency. Constriction of bronchi and bronchioles in chronic allergic bronchitis and bronchiolitis results in similar clinical signs. However, administration of bronchodilators results in rapid relief of airway obstruction in cases of allergic bronchitis (eg, heaves in horses). Some chronic pneumonias (eg, maedi in sheep) are characterized by firm diffuse lesions due to hyperplasia of lymphoid follicles, hyperplasia of smooth muscle around bronchioles, diffuse fibrosis, and diffuse lymphocytic infiltration. Aspiration pneumonia often leads to gangrene, with severe toxemia accompanying the acute inflammatory reaction.

Most infectious pneumonias develop in the anteroventral portions of the lungs. However, infectious agents, as well as neo-plasms, can invade the lungs via the blood, which may extensively impair pulmonary function, as can pulmonary edema from chronic heart failure. Pleuritis, empyema, hydrothorax, chylothorax, atelectasis, diaphragmatic hernia, or pneumothorax can also seriously impair respiratory function. Pulmonary thrombosis leads to acute, often fulminant, respiratory failure as a result of a lack of pulmonary arterial blood flow to ventilated regions of the lung. Infarction of the lung can reduce respiratory function but is rare because of the dual blood supply of the organ. Toxic injury, such as in 3-methylindole toxicity in cattle, causes edema, emphysema, and necrosis of alveolar epithelium, followed by compensatory hyperplasia of these cells; the effects on gas exchange result in severe hypoxia and dyspnea. Although pneumonia is most important, several other thoracic conditions can cause respiratory dysfunction. Pulmonary edema, the abnormal accumulation of fluid in the interstitial tissue, airways, or alveoli of the lungs, may occur in conjunction with circulatory disorders, particularly left ventricular failure or increased capillary permeability, occasionally in anaphylactic and allergic reactions, and in some infectious diseases. Head trauma can cause pulmonary edema in dogs. Dyspnea and open-mouth breathing may occur. Animals stand in preference to lying down, lie only in sternal recumbency, or may assume a sitting position. Auscultation of the chest may reveal wheezing and fluid sounds.

Pleuritis (pleurisy) may be caused by any pathogen that gains entrance to the pleural cavity, but it is often an extension of pneumonia. Rapid shallow breathing, fever, and thoracic pain are suggestive of pleuritis. Auscultation of the chest may reveal friction sounds.

Empyema (purulent exudate in the pleural cavity) is caused by pyogenic bacteria or fungi reaching the thoracic cavity via the blood or by extension of a pneumonia, traumatic reticulitis, or penetrating wound of the chest. Cough, fever, pain, and dyspnea may be present.

Hemothorax (the accumulation of blood in the pleural cavity) is usually caused by trauma to the thorax, systemic coagulopathy, or thoracic neo-plasia. Hydrothorax (the accumulation of transudate in the pleural cavity) is usually due to interference with venous blood flow or lymph drainage. Chylothorax (the accumulation of chyle in the pleural cavity) is relatively rare and is seen most often in cats. It may be caused by rupture of the thoracic duct but often is idiopathic. The signs of all three conditions include respiratory embarrassment (eg, rapid shallow breathing with inspiratory dyspnea) and weakness.

Pneumothorax (air in the pleural cavity, see Treatment) may be of traumatic or spontaneous origin. Air can enter the pleural cavity through penetrating wounds of the thoracic wall or by extension from pulmonary emphysema or ruptured bullae. The lung collapses if a large volume of air enters the pleural cavity. Bilateral pneumothorax may develop if the mediastinum is weak or incomplete. Inspiratory dyspnea or rapid, shallow breathing is evident.

RESPIRATORY DISEASE AND ITS CONTROL

Sudden dietary changes, weaning, cold, drafts, dampness, dust, high levels of ammonia, poor ventilation in general, and the mixing of widely divergent age groups all play a role in respiratory disease in groups of animals.

Stress and mixing of animals from several sources should be avoided or minimized. Establishing individual animal identification, making accurate clinical and postmortem diagnoses, and maintaining a record system of diagnosis and treatment are important to minimize or control outbreaks of pneumonia. Transportation over long distances is another stress factor that plays a major role in the pathogenesis of respiratory infections in large animals. Immunization can help control respiratory infection. However, control may be compromised by improper timing, use of ineffective or inappropriate vaccines, or overwhelmingly negative management practices. In most cases, severe insults to the natural defences cannot be reversed later by therapeutic agents and biologicals.

The mucosal surfaces of the respiratory tract contain lymphoid follicles that exchange cells with other parts of the body. However, most of the lymphocytes in the respiratory lining produce only IgA, whereas the cells in the lymph nodes of the respiratory tract produce IgM and IgG. Depending on the agent involved, various cell- and antibody-mediated immune responses occur in the respiratory tract and include opsonization, agglutination, immobilization, neutralization of toxins and viruses, blockage of adherence to cells, lysis, and chemotaxis. The type of immune response varies because of age, species, and the means to respond to specific virulence mechanisms of the pathogens involved.

Species vary in the type of immune response available at different sites in the respiratory tract. Large antigen droplets may immunize the upper tract with IgA, but small replicating particles may be necessary to immunize the lower tract. To develop adequate antibody levels to protect the lungs, repeated doses of antigen plus adjuvant, or a replicating antigen, are often necessary. These results are seldom achieved under field conditions (eg, many field trials using respiratory vaccines in cattle have not demonstrated statistically significant efficacy).

Environmental management is an essential part of therapy in allergic respiratory diseases. For example, clinical signs in horses with heaves or cattle with hypersensitivity pneumonitis may be effectively controlled by preventing exposure to molds present in hay.

RESPIRATORY DISEASE AND PRINCIPLES OF THERAPY

Respiratory disease is often characterized by abnormal production of secretions and exudates and by a reduced ability to remove them. The primary goal of therapy is to reduce the volume and viscosity of the secretions and to facilitate their removal.

This can be accomplished by controlling infection and inflammation, modifying the secretions, and when possible, improving postural drainage and mechanically removing the material. Therapeutic methods include altering the inspired air and administering expectorants, antitussives, bronchodilators, antimicrobials, diuretics, and other drugs.

However, expectorants have shown little or no beneficial effects in clinical trials. Hydration should be maintained.

Inhalation of humidified air may facilitate removal of airway secretions. Expectorants are sometimes used with the intention of liquefying these secretions.

However, they should be used in conjunction with ancillary respiratory therapy such as improved postural drainage, mild exercise, and thoracic percussion, which (in addition to coughing) encourages expectoration and removal of secretions. Mechanical removal of tenacious and viscid secretions by aspiration may be necessary in severe airway obstruction.

Antitussive agents are indicated to relieve the discomfort associated with non-productive coughing but are contraindicated when secretion of airway mucus is excessive. Products that contain atropine also are contraindicated, at least in theory, because atropine increases the viscosity of airway secretions.

Increased airway resistance caused by bronchial smooth muscle contraction can be alleviated with bronchodilators, which may be indicated in animals with asthma-like conditions and chronic respiratory disease.

Methylxanthines, such as theophylline and aminophylline, are effective bronchodilators in species other than cattle (and possibly dogs); however, the therapeutic index is relatively narrow and they are less efficacious than β_2-agonists. Isoproterenol, clenbuterol, and epinephrine are also generally effective, and sodium cromoglycate may be used in horses with inflammatory airway disease.

Corticosteroids are highly effective in allergic conditions, but systemic use may result in adverse effects. Aerosolized corticosteroids are efficacious and associated with few to no adverse effects; however, they require an aerosol delivery device (eg, face mask) for proper administration. Antihistamines can be used to alleviate the bronchoconstriction caused by histamine release; however, they are of limited value in large animals. Bronchospasm also can be reduced significantly by removing irritating factors, using mild sedatives, or reducing periods of excitement.

In bacterial infection, antimicrobial therapy should be instituted. The goal is to select either the most effective agent against a specific organism or the least toxic agent of several alternatives. Culture and sensitivity testing of airway secretions provide a worthwhile, although not infallible, guide to determining the appropriate antibiotic. Knowledge of tissue penetration and pharmacokinetic characteristics of the antimicrobial agents is important as well.

The agents have proved effective in the listed species: cattle-oxytetracycline, cephalosporins, fluoroquinolones, macrolides, florfenicol, penicillins, and sulfonamides; sheep and goats-oxytetracycline, cephalosporins, macrolides, penicillins, and sulfonamides; pigs-lincomycin, spectinomycin, penicillins, and sulfonamides; dogs and cats cephalosporins, chloramphenicol, amoxicillin-clavulanate, aminoglycosides, trimethoprim-sulfamethoxazole, fluoroquinolones, macrolides, and tetracyclines; horses-penicillins, aminoglycosides, cephalosporins, fluoroquinolones, sulfonamides, and tetracyclines (the latter with caution due to an occasional adverse effect of severe diarrhea).

Aminoglycosides are useful but can be nephrotoxic. Trimethoprim, usually in combination with a sulfonamide, is useful for respiratory therapy in most species but is not licensed for food-producing animals in the USA. Drugs such as enrofloxacin (approved for small animals and cattle but not for horses in the USA) and ceftiofur are effective for pneumonia.

Broad-spectrum antibiotics should be used if specific bacteria cannot be identified, and once begun, a full course of therapy should be completed. Multiple antimicrobial agents should be used only with full knowledge of the potential drug interactions.

Because of residues in food-producing animals, veterinarians must use these products according to label instructions and provide sound advice to producers. Extra-label use of antimicrobials is permitted in some situations and is regulated by the Animal Medicinal Drug Use Clarification Act of 1994.

The hypoxemia caused by most lung disorders usually can be corrected by administering oxygen. However, continuous administration of high concentrations increases the tendency for regional resorption atelectasis, thus worsening the hypoxemia, and can cause pneumonitis on its own. Hypoxemia is often accompanied by variable degrees of hypercapnia and acidemia. Endotracheal intubation and mechanical ventilation may be necessary in animals with acute respiratory failure or in animals that are comatose or apneic. Arterial blood gas and pH determinations, when practicable, are extremely valuable to monitor treatment.

Diuretics are indicated in pulmonary edema. The osmotic diuretics have a minimal action on diuresis, carbonic anhydrase inhibitors (eg, acetazolamide) have a moderate effect, and loop diuretics (eg, furosemide) have a profound effect.

6

Animal Morphology and Anatomy

Animal Morphology or anatomy, is the study of the shape, form, and structure of animals and their parts. Morphologists are interested in questions such as how the fang mechanisms of cobras differ from those of rattlesnakes or how the wings of bats evolved from the forelimbs of their ancestors.

Animals can be broadly distinguished into two major groups the nonchordates (or invertebrates) and the chordates.Invertebrates account for more than 95% of the animal species that have been identified. These are animals which do not have a vertebral column or a back bone. They are distributed into nine major phyla from protozoans to echinoderms. The chordates are animals which have a notochord on the dorsal side of the body. Most of the chordates can be described as vertebrates since the notochord in these animals is restricted to the embryonic stage and in the adults it becomes replaced by a vertebral column.

A very distinct evolutionary trend in the general organisation of animals can be noticed as one proceeds from protozoans to chordates. From an acellular level of organisation to cellular level and then tissue level, organ level and organ system level of organization can be recognised. In addition, one can recognise the gradual appearance of distinct features such as symmetry, body wall and body cavity. There are further improvements in the form of appearance of segmentation, body divisions and even formation of a distinct head (cephalistion).

The culmination of all these evolutionary advances, is seen in the mammals which represent the most recent group of animals to have evolved and at the same time the most advanced group of animals. In order to gain an insight into the evolutionary advances among animals, it become necessary to study the salient features of representative animals from the group of invertebrates and vertebrates.

If one examines a living cell under the highest powers of the microscope one sees surprisingly little. In view of the fact that protoplasm is capable of complicated physical and chemical transformations it might be supposed that it would possess intricate mechanical arrangements. As everyone knows, the visible structure of a living cell is rather simple, and there is nothing to suggest the complexities of man-made machines. This may mean that the fundamental structure of protoplasm is ultramicroscopic or in the range of molecular dimensions. Following the establishment of the cell doctrine the science of cytology was built up mostly along morphological lines.

Cytologists sought to find in the structure of the cell some explanation of its life. In their study of the nucleus they were to some extent successful. Some years ago, the study of the huge chromosomes found in the cells of the salivary glands of fly larvae completely confirmed the earlier work of T. H. Morgan and his school, and showed that the genes are arranged in linear fashion along the chromosomes.

Thus the visible structure of the chromosome bears a relation to its function in heredity. The position of the genes is important in determining the way in which they are distributed under various conditions, but the fact that they are arranged in linear order tells us nothing as to how a gene can influence the development of one character or another. In their study of the structure of cytoplasm cytologists have not been so fortunate.

Although the cell nucleus is of primary importance in inheritance, for the ordinary routine life of the cell the cytoplasm plays the leading role. This is indicated by various experiments in which cells were deprived of nuclei. Thus, Lynch, among others, divided amebae into halves, one with nucleus, the other without. The fragment without a nucleus was able to live for days, although apparently it was incapable of synthesizing nitrogenous compounds. Similar studies of other workers indicate that cells without nuclei can live for a time.

A surprising case is that of Acetabularia mediterranea, a large unicellular alga which possesses but a single nucleus that is easy to remove. In this form, cytoplasm without nucleus can live for months and is capable of regenerating some rather complicated structures. On the other hand, in the nerve cells of higher animals, growth of cytoplasm can occur only in the region of the nucleus. Studies of the structure of cytoplasm have yielded very little which has proved of much help in the interpretation of its function.

Perhaps because of this, some modern cytologists seem to be losing interest in the classical cytology which dealt almost entirely with structure and have become more and more concerned with the chemistry and the physiology of the cell. All living cells are believed to be surrounded by definite membranes. The evidence for this statement has come almost entirely from the physiological side. Since the early days of cytology morphologists have been but little concerned with the outer boundaries of cells. If one examines a sea urchin egg carefully one can see that it is bounded by a delicate membrane perhaps 1μ in thickness.

This membrane is very hard to distinguish, for it appears to have the same refractive index as the sea water surrounding the egg. Whether the membrane which is visible around the sea urchin egg is the same membrane as that which governs the osmotic behaviour of the cell is not certain. The simplest theory would be to assume that it is. However, some authors insist that osmotic membranes surrounding cells are much thinner than the membrane which can be seen to envelop the sea urchin egg.

It is possible that the cell boundaries or cell walls which can be so clearly seen in preparations of animal tissues are really membranes of physiological importance. The question as to whether they are or not has scarcely been discussed. Most biologists are influenced by the relation one finds in plant tissues.

The cell walls of plant cells are non-living, and serve merely as supports for the protoplasmic structures within. On being placed in concentrated solutions plant cells shrink away from their walls. There is evidence that, within the outer membrane of the animal cell, there is a cortex, different both in morphological and physical properties from the cell interior.

In some protozoan cells such a cortex can be clearly recognized. In metazoan cells it is not so easy to distinguish. However, Just has presented both morphological and physiological evidence to show the presence of a cortex as a very important part of the metazoan cell. Within the outer membrane and the cortex of a living cell lies the main bulk of the cytoplasm. Cytologists have struggled hard to discover structures in the cytoplasm which would help to explain its activity. Because of the fact that living cells reveal relatively little, various methods of fixing, sectioning, and staining have been employed.

Cells, or groups of cells, are treated with strong acids, salts of heavy metals, alcohol or formaldehyde, or with various combinations of these and other reagents. Following fixation the cells are bathed in alcohol, then in some solvent of paraffin, and then (typically) they are impregnated with paraffin, cut into sections, and stained. These methods of treatment have been very successful in the study of chromosomes, and it can be shown by comparison with the living that the chromosomes are not markedly altered by all these procedures. The value of fixation and staining in the study of cytoplasm is less certain.

Fischer showed at great length that fixation agents caused the production of various artifacts in protein solutions, and that these artifacts resembled the granules and filaments described by cytologists as occurring in living cells. Indeed, if one watches the action of a fixing agent on living cells, one may, under proper conditions, observe the formation of new granules or vacuoles, or even the disappearance of granules normally present.

Since the publication of Fischer's book cytologists have been more careful to compare the structures they see in fixed and stained material with the living protoplasm. In the latter half of the nineteenth century students of the cell urged one type of structure or another as essential for the living protoplasm.

Thus, Frommann, Heitzmann, Klein and Leydig, all argued that protoplasm consisted of a net of solid material, the meshes of which were filled with fluid. Flemming believed that fibrils were the most important constituent of the protoplasm. On the other hand, Altmann vigourously maintained the importance of granules. He believed that the granules represented minute organisms living within a more or less inanimate fluid.

Finally, the protozoologist, Bütschli, thought of protoplasm as a foam, or rather as an emulsion composed of 2 liquids, one in the form of droplets, the other as lamellae between the droplets. These nineteenth century views of the fundamental nature of living substance are not as new as is generally supposed.

Even in the eighteenth century many observers had noted the presence of innumerable granules in living material, and there was also some talk of the existence of fibrils. The twentieth century saw a new development, or rather a variation of the old one. In 1897, Benda coined the word *mitochondria* for granules which he found first in rat spermatocytes and later in other types of cells. These mitochondria often showed a tendency to unite in the form of threads. Benda believed that his complicated technic was specific for the mitochondria, and that these represented a new discovery.

Actually, mitochondria are not to be distinguished from the granules of Altmann or the filaments of Flemming. It is important to note that not all the granules of the cytoplasm are to be considered as mitochondria. Nor is it always entirely clear just which granules are mitochondria and which ones are not. In general mitochondria are coloured with the vital dye Janus green, and Sorokin regards Janus green B as "virtually specific" for mitochondria. They can be demonstrated by the use of those fixatives which do not dissolve lipids, and it is believed that chemically mitochondria consist of a mixture of protein with a phospholipid (e.g., lecithin).

Mitochondria can be seen in living protoplasm, and they are believed to be necessary constituents of all plant and animal cells. In plant cells mitochondria may become transformed into plastids. In animal cells they have been claimed to transform into secretion granules, fibrils, and various other structures of differentiated cells.

Concerning their physiological function little has been known until recently. Now it is believed that mitochondria contain oxidizing enzymes, especially the enzymes known as succinoxidase and cytochrome oxidase. Soon after the discovery of mitochondria, or rather soon after the introduction of the concept by Benda, another type of cytoplasmic structure was introduced to the cytological world.

Studying the nerve cells of an owl, Golgi described the presence of a network, the so-called "Golgi apparatus." This may be in the form of a net, or it may be in the form of granules or rods. It is demonstrated either by a silver impregnation process or by long continued soaking in osmic acid (OsO_4) solutions. For many years the Golgi apparatus was believed to have a great significance, for it could be demonstrated in many diverse types of protoplasm; and also, according to a number of authors, it could be seen in various types of living cells. Some workers have found that the region of the cell which contains the Golgi apparatus is rich in phosphatase enzymes, that is, enzymes which are able to split phosphate from organic combination.

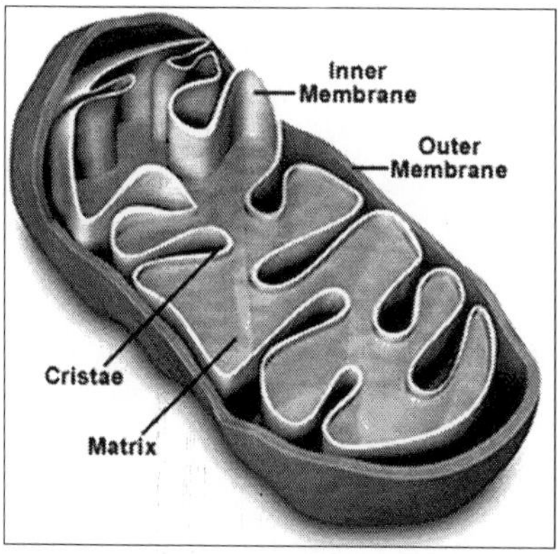

Fig. ***Mitochondria***

However, in recent years there has been growing scepticism concerning the x. According to Hibbard, a leading worker in the field, "there does not seem to be much agreement as to the nature and functioning of the Golgi apparatus" and she feels that in so far as most of the generalizations that have been made are concerned, "we should be better off if the discovery of the impregnation methods had never been made."

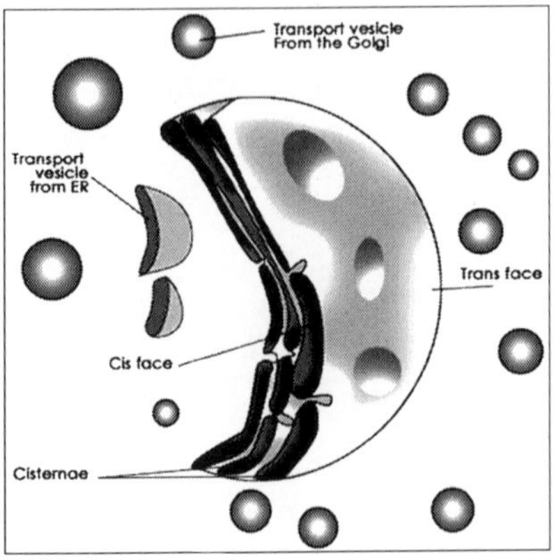

Fig. ***Golgi Apparatus***

From time to time, it has been urged that the Golgi apparatus as such does not exist in living cells at all, but is an artifact produced by the reagents which are used to demonstrate it. Palade and Claude have presented forcible arguments in support of this view, but these arguments have been vigourously disputed.

Finally, many cells contain vacuoles, which may be minute or large. A comprehensive term for the vacuoles of a cell is "vacuome." Vacuoles may appear and disappear. They can originate from mitochondria, or be quite distinct from them. Sometimes they are produced by reagents, as will be noted later.

Occasionally cells are so filled with vacuoles that the protoplasm has the appearance of a foam. This type of protoplasm is most common in the protozoa, and it is easy therefore to understand why Bütschli regarded all protoplasm as foamlike. Students of the finer structures of cells have always attempted to use microscopes which show the greatest possible detail. For this purpose, the limiting factor is not the extent to which the microscopic image may be enlarged, for the degree of possible enlargement is infinite.

The limiting factor is the "resolving power," or the power of discriminating between 2 minute objects that lie close together. Years ago, the resolving power of ordinary microscopes was developed to its very greatest extent, for the resolving power is primarily dependent on the wave length of the light passing through the lenses of the microscope and is actually inversely proportional to this wave length. By the use of ultraviolet radiation, which has shorter wave lengths than visible light, a 2 1/2 fold increase in resolving power could be obtained, and ultraviolet microscopes were developed. In these, quartz lenses were employed, inasmuch as glass is opaque to ultraviolet radiation. A far greater increase in resolving power may be obtained if instead of visible light or ultraviolet radiation, cathode rays (beams of electrons) are employed. Modern physics regards the electron as a particle having a wave nature, and the wave length of such radiation may be only 1/100,000th that of visible light.

Hence it might be possible to obtain vastly greater resolving power than that possible with the ordinary microscope. Developments in the new science of electron optics have made possible the construction of electron microscopes. In such microscopes the electrons are focused by electron lenses, which consist of electrostatic or magnetic fields capable of deflecting the electrons.

Cathode rays are effective only in a vacuum, and biological material to be studied in an electron microscope must be thoroughly dried. The drying process itself can cause violent distortion. Thus if some cells are suspended in a droplet of water and this is then allowed to evaporate, the surface of the water eventually encounters the cells. Surface tension exerts relatively enormous forces on small objects and the cells may be badly crushed.

To avoid such surface effects, Anderson has found it desirable to replace the water with liquid carbonic acid which is gaseous at ordinary temperatures, and then to evaporate at temperatures above the critical temperature of the gas. In this way, better pictures can be obtained.

Of course the electrons or cathode rays themselves are destructive so that living material even if it were to survive the drying would be killed by the cathode rays. In spite of all these shortcomings, the electron microscope has been able to demonstrate within bacteria structures not visible with the ordinary microscope, and it is extremely useful in the study of bacteriophages and viruses.

An electron microscope picture of bacteria and bacteriophages is shown in figure. In order to examine larger cells under the electron microscope they must either be pulled apart, which can cause violent change, or they must be studied in thin sections.

For the preparation of very thin sections, special microtomes have been constructed. So far the study of large cells with the electron microscope has not yielded results of any great interest or importance. The morphological study of the cytoplasm has resulted in a large body of information concerning the various granular and other formed constituents of living substance. Unfortunately, however, there are no very definite clues as to the relation of these formed elements to the vital activity. One point seems certain: the granules or filaments of the cytoplasm are not present in any spatial arrangement of significance for the physiology of the cell.

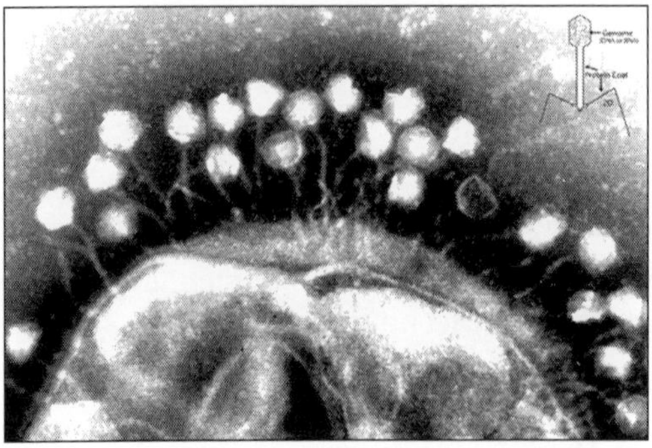

Fig. Bacteriophage

At least this is true for the main mass of the protoplasm, although it may not be true for the cortex. When cells are centrifuged, even at very high speeds, the distribution of the granules may be profoundly altered without affecting the life or subsequent behaviour of the cell.

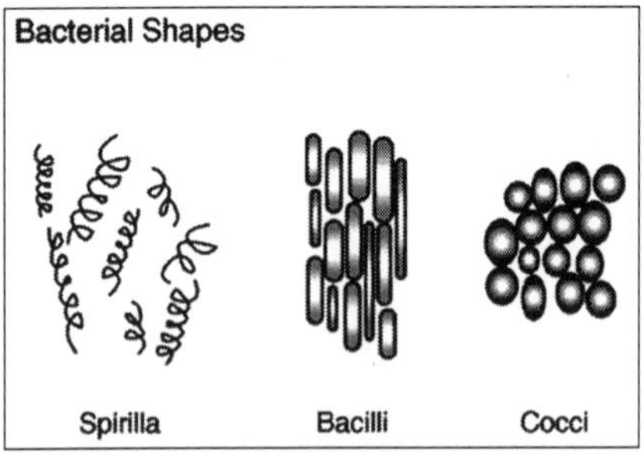

Fig. ***Bacterial Shapes***

It should be noted that the granules of living cells may be present in tremendous numbers. In the egg of the sea urchin Arbacia, the granules occupy nearly 20 per cent of the total volume. In the egg of the starfish Asterias there is a much higher concentration of granules, and the egg of the sea urchin Paracentrotus is so full of granules that there is hardly any free space between them.

KINGDOM

The sub-kingdom Vertebrata may be shortly defined as comprising animals in which the body is composed of a number of definite segments arranged along a longitudinal axis; the nervous system is in its main masses dorsal, and the neural and haemal regions of the body are always completely shut off from one another by a partition; the limbs are never more than four in number, and are always turned away from the neural aspect of the body; mostly there is the bony axis known as the "spine" or "vertebral column" and in all the structure known as the "noto-chord" is present — in the embryo, at any rate.

These characters distinguish the Vertebrata, as a whole, from the Invertebrata; but it is necessary to define these broad differences more minutely, and to consider others which are of little less importance.

One of the most obvious, as it is one of the most fundamental, of the distinctive characters of Vertebrates, is to be found in the shutting off of the main masses of the nervous system from the general cavity of the body. In all Invertebrate animals, without exception, the body (figure A) may be regarded as a single tube, enclosing all the viscera; and consequently, in this case, the nervous system is contained within the general cavity of the body, and is not in any way shut off from the alimentary canal.

The transverse section, however, of a Vertebrate animal exhibits two tubes (figure B), one of which contains the great masses of the nervous system - that is, the cerebro-spinal axis, or brain and spinal cord - whilst the other contains the alimentary canal and the chief circulatory organs, together with certain portions of the nervous system known as the "ganglionic" or "sympathetic" system.

Leaving the cerebro - spinal centres out of sight for a moment, we see that the larger or visceral tube of a Vertebrate animal contains the digestive canal, the haemal system, and a gangliated nervous system. Now this is exactly what is contained in the visceral cavity of any of the higher Invertebrate animals; and the opinion has been generally entertained that it is the sympathetic nervous system of Vertebrates which is truly comparable to, and homologous with, the nervous system of Invertebrates. On the other hand, there are Invertebrates with a sympathetic system of nerves, and the development of the nerve-chain of the Annu-losa resembles that of the cerebro-spinal axis of the Vertebrata. The tube containing the cerebro-spinal centres is formed as follows: At an early period in the development of the embryo of any Vertebrate animal, the portion of the ovum in which development is going on - the "germinal area" - becomes elevated into two parallel ridges, one on each side of the middle line, enclosing between them a long groove, which is known as the " primitive groove".

The ridges which bound the primitive groove are known as the " laminae dorsales;" and they become more and more raised up, till they ultimately meet in the middle line, and unite to form a tube, within which the cerebro-spinal nervous centres are developed. It follows from its mode of formation that the inner wall of the tube formed by the primitive groove, which remains as the septum between the cerebro-spinal canal and the body-cavity, is nothing more than a portion of the primitive wall of the body of the embryo. And there appears to be little doubt, as believed by Remak and Huxley, that the cerebro-spinal nervous centres are "the result of a modification of that serous layer of the germ, which is continuous elsewhere with the epidermis ".

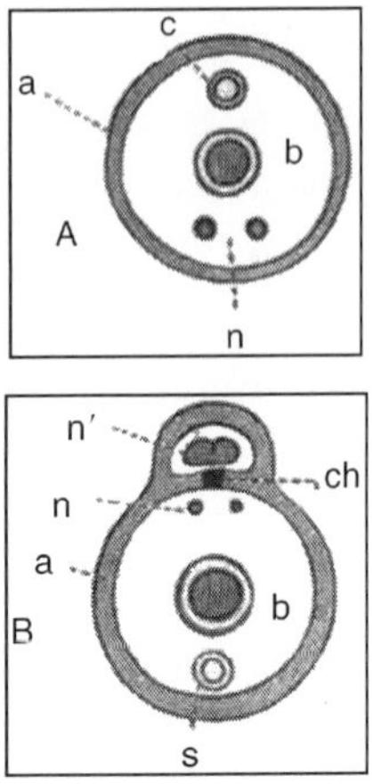

Fig. A, Transverse Section of the Body of one of the Higher Invertebrata:

a. Body-wall;
b. Alimentary Canal;
c. Haemal System;
n. Nervous System.
B. Transverse Section of the Body of a Vertebrate Animal:
a. Body-wall;
b. Alimentary Canal;
c. Haemal System;
n. Sympathetic System of Nerves;
n. Cerebro-spinal System of Nerves;
ch. Notochord.

Another remarkable peculiarity as regards the nervous system is found in the fact that in no Vertebrate animal does the alimentary canal pierce the main masses of the nervous system, but turns away to open on the opposite side of the body. In most Invertebrates, on the other hand, in which there is a well-developed nervous system, this is perforated by the gullet, so that an oesophageal nerve-collar is formed, and some of the nervous centres become prae — oesophageal, whilst others are post-oesophageal.

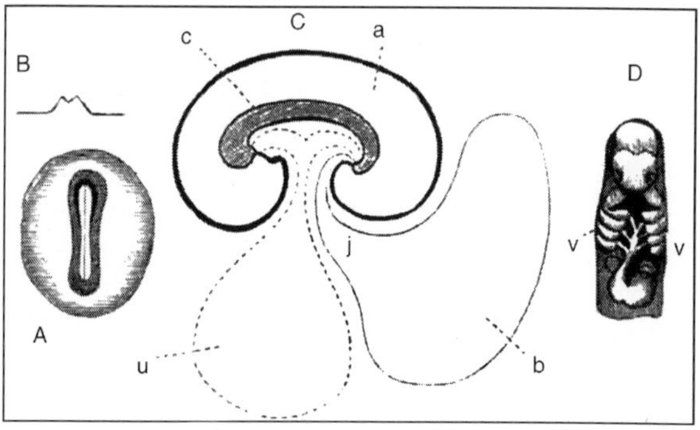

The sub-kingdom Vertebrata may be shortly defined as comprising animals in which the body is composed of a number of definite segments arranged along a longitudinal axis; the nervous system is in its main masses dorsal, and the neural and haemal regions of the body are always completely shut off from one another by a partition; the limbs are never more than four in number, and are always turned away from the neural aspect of the body; mostly there is the bony axis known as the "spine" or "vertebral column" and in all the structure known as the "noto-chord" is present - in the embryo, at any rate. These characters distinguish the Vertebrata, as a whole, from the Invertebrata; but it is necessary to define these broad differences more minutely, and to consider others which are of little less importance.

One of the most obvious, as it is one of the most fundamental, of the distinctive characters of Vertebrates, is to be found in the shutting off of the main masses of the nervous system from the general cavity of the body. In all Invertebrate animals, without exception, the body (figure,) may be regarded as a single tube, enclosing all the viscera; and consequently, in this case, the nervous system is contained within the general cavity of the body, and is not in any way shut off from the alimentary canal.

The transverse section, however, of a Vertebrate animal exhibits two tubes (figure,), one of which contains the great masses of the nervous system - that is, the cerebro-spinal axis, or brain and spinal cord - whilst the other contains the alimentary canal and the chief circulatory organs, together with certain portions of the nervous system known as the "ganglionic" or "sympathetic" system. Leaving the cerebro - spinal centres out of sight for a moment, we see that the larger or visceral tube of a Vertebrate animal contains the digestive canal, the haemal system, and a gangliated nervous system. Now this is exactly what is contained in the visceral cavity of any of the higher Invertebrate animals; and the opinion has been generally entertained that it is the sympathetic nervous system of Vertebrates which is truly comparable to, and homologous with, the nervous system of Invertebrates.

On the other hand, there are Invertebrates with a sympathetic system of nerves, and the development of the nerve-chain of the Annu-losa resembles that of the cerebro-spinal axis of the Vertebrata. The tube containing the cerebro-spinal centres is formed as follows: At an early period in the development of the embryo of any Vertebrate animal, the portion of the ovum in which development is going on - the "germinal area" - becomes elevated into two parallel ridges, one on each side of the middle line, enclosing between them a long groove, which is known as the " primitive groove". In some Vertebrates, such as the Lancelet (Amphioxus), the notochord is persistent throughout life. In the majority of cases, however, the notochord is replaced before maturity by the structure known as the "vertebral column" or "backbone," from which the sub-kingdom Vertebrata originally derived its name. This is not the place for an anatomical description of the spinal column, and it is sufficient to state here that it is essentially composed of a series of cartilaginous, or more or less completely ossified, segments or vertebra, arranged so as to form a longitudinal axis, which protects the great masses of the nervous system. It is to be remembered, however, that all Vertebrate animals do not possess a vertebral column. They all possess a notochord; but this may be persistent, and in many cases the development of the spinal column is extremely imperfect.

Another embryonic structure which is characteristic of all Vertebrates, is found in the so-called " visceral arches" and "clefts". The "visceral arches" are a series of parallel ridges running transversely to the axis of the body, situated at the sides of, and posterior to, the mouth. As development proceeds, the intervals between these ridges become grooved by depressions which gradually deepen, until they become converted into a series of openings or "clefts," whereby a free communication is established between the upper part of the alimentary canal (pharynx) and the external medium. In Fishes and many Amphibians the greater number of the visceral clefts remain open throughout life; and the visceral arches of all fishes (except the Lancelet) throw out filamentous or lamellar processes, which receive branches of the aorta and constitute branchiae. In the higher Vertebrata all the visceral clefts become closed, whilst no branchiae are ever developed upon the visceral arches.

The limbs of Vertebrate animals are always articulated to the body, and they are always turned away from the neural aspect of the body. They may be altogether wanting, or they may be partially undeveloped; but there are never more than two pairs, and they always have an internal skeleton for the attachment of the muscles of the limb. A specialised blood-vascular or "haemal" system is present in all the Vertebrata; and in all except one - the Amphioxus

- there is a contractile cavity or heart, which never consists of less than two chambers provided with valvular apertures. In all the Vertebrata the heart is essentially a respiratory heart - that is to say, it is concerned with driving the impure or venous blood to the breathing organs; and in its simplest form (fishes) it is nothing more than this. In the higher Vertebrates, however, there is superadded to this a pair of cavities which are concerned in driving the pure or arterial blood to the body. In the case of the Mammals, these two circulations are often spoken of as the "lesser" or "pulmonary" circulation, and the "greater" or "systemic" circulation.

In all Vertebrates there is that peculiar modification of the venous system which is known as the "hepatic portal system." That is to say, a portion of the blood which is sent to the alimentary canal, instead of returning to the heart by the ordinary veins, is carried to the liver by a special vessel - the vena portae - which ramifies through this organ after the manner of an artery.

In all Vertebrates, also, is found the peculiar system of vessels known as the " lacteal system." This is to be regarded as an appendage of the venous system of blood vessels, and consists of a series of vessels which take up the products of digestion from the alimentary canal, elaborate them, and finally empty their contents into the veins.

Lastly, the masticatory organs of Vertebrates are modified portions of the walls of the head, and never " hard productions of the alimentary mucous membrane, or modified limbs ", as they are amongst the Invertebrata.

The above are the leading characters of the Vertebrata as a whole; but before going on to consider the primary divisions of the sub-kingdom, it may be as well to give a very brief and general description of the anatomy of the higher and more typical Vertebrates, commencing with their bony framework, or skeleton. The skeleton of the Vertebrata may be regarded as consisting essentially of the bones which go to form the head and trunk on the one hand (sometimes called the " axial" skeleton), and of those which form the supports for the limbs (" appendicular " skeleton) on the other hand. The bones of the head and trunk may be looked upon as essentially composed of a series of bony rings or segments, arranged longitudinally, one behind the other. Anteriorly these segments are much expanded, and likewise much modified, to form the bony case which encloses the brain, and which is termed the cranium or skull.

Behind the head the segments enclose a much smaller cavity, which is called the "neural" or spinal canal, as it encloses the spinal cord; and they are arranged one behind the other, forming the vertebral column. The segments which form the vertebral column are called "vertebrae," and they have the following general structure: Each vertebra (figure, A) consists of a central piece, which is the fundamental and essential element of the vertebra, and is known as the "body" or "centrum" (V). From the upper or posterior surface of the centrum spring two bony arches (n n), which are called the "neural arches" or "neurapophyses," because they form with the body a canal - the "neural canal" - which encloses the spinal cord. From the point where the neural arches meet behind, there is usually developed a longer or shorter spine, which is termed the " spinous process," or "neural spine" (s). From the neural arches there are also developed in the typical vertebra two processes (a a), which are known as the "articular " processes, or "zyga-pophyses."

DESCRIPTION

The vertebrae are united to one another partly by these, but to a greater extent by the bodies or "centra." From the sides of the vertebral body, at the point of junction with the neural arches, there proceed two lateral processes (d d), which are known as the "transverse processes." (In the typical vertebra the transverse processes consist each of two pieces, an anterior piece or "parapophysis," and a posterior piece or "diapophysis.") These elements form the vertebra of the human anatomist, but the "vertebra" of the transcendental anatomist is completed by a second arch which is placed beneath the body of the vertebra, and which is called the "haemal" arch, as it includes and protects the main organs of the circulation. This second arch is often only recognisable with great difficulty, as its parts are generally much modified, but a good example may be obtained in the human chest, or in the caudal vertebra of a bony fish.

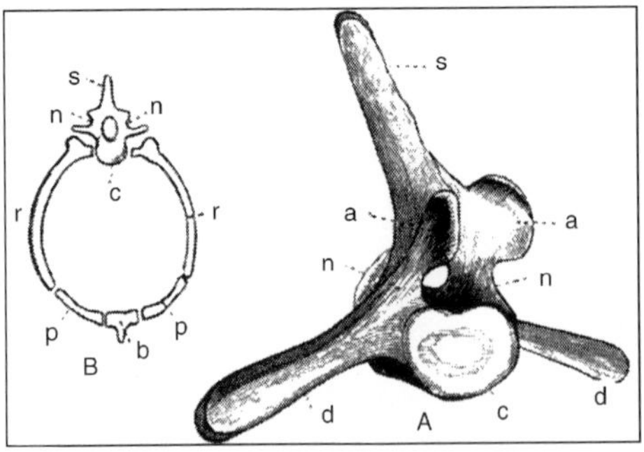

Fig. A, Lumbar Vertebra of a Whale: c Body or Centrum; n n Neural Arches; s Neural Spine; a a Articular Processes; d d Transverse Processes. B, Diagram of a Thoracic Vertebra: c Centrum; n n Neural Arches Enclosing the Neural Canal; s Neural Spine; r r Ribs, Assisting in the Formation of the Haemal Arch; p p Costal Cartilages; b Sternum, with Haemal Spine.

The haemal arch in the case of the human thorax is formed by the ribs (r r) and the costal cartilages (p p), and is completed in front by the breast-bone or sternum (b), which in some cases - but not in man - develops a spine (the haemal spine) which corresponds to the neural spine on the opposite aspect of the vertebra.

It follows from the above, that the typical vertebra consists of a central piece or body from which two arches are given off, one of which protects the great masses of the nervous system, and is therefore said to be "neural;" whilst the other protects the main organs of the circulation, and is therefore said to be "haemal." The correspondence of the typical bony segment or vertebra with the doubly tubular structure of the body in all Vertebrates is thus too obvious to require to be specially pointed out.

As a general rule, the vertebral column is divisible into a number of distinct regions, of which the following are recognisable in man and in the higher Vertebrata:

- A series of vertebrae which compose the neck, and constitute the "cervical region" of the spine.
- A number of vertebrae which usually carry well-developed ribs, and form the "dorsal region" (d).
- A series of vertebrae which form the region of the loins, or "lumbar region"(l).
- A greater or less number of vertebrae Which constitute the "sacral region," and are usually amalgamated or "anchylosed" together to form a single bone, the "sacrum."
- The spinal column is completed by a variable number of vertebrae which constitute the "caudal" region, or tail (c).

As regards the skull of the Vertebrata, it has been thought advisable not to enter into any general details here, partly because the subject is one which can only be properly discussed in a work specially devoted to Human or Comparative Anatomy, and partly because there is still much diversity of opinion as to the exact composition of the skull. There is, however, a very general concurrence of opinion that the skull is composed of a series of separate segments, and this is a point which it is important to remember. By Owen, and by many other competent authorities, these cranial segments are looked upon as being nothing more than so many vertebrae, the neural canals of which are greatly expanded to enclose the brain, whilst the haemal arches are very greatly modified to serve different purposes.

This view is not accepted by high authorities; but the general fact that the skull is composed of separate segments is universally admitted. The only portion of the bony framework of the head which it is absolutely essential to understand, is the lower jaw or "mandible." The lower jaw is sometimes wanting, but when present, it consists in all Vertebrata of two halves or "rami," which are united to one another in front, and articulate separately with the skull behind. In many cases, each half, or "ramus," of the lower jaw consists of several pieces united to one another by sutures; but in the Mammalia each ramus consists of no more than a single piece. The two rami are very variously connected with one another, being sometimes only joined by ligaments and muscles, sometimes united by cartilage or by bony suture, and sometimes fused or anchylosed with one another, so as to leave no evidence of their true composition. The mode by which each ramus of the lower jaw articulates with the skull also varies. In the Mammalia the lower jaw articulates with a cavity formed on what is known to human anatomists as the temporal bone; but in Birds and Reptiles the lower jaw articulates with the skull, not directly, but by the intervention of a special bone, known as the "quadrate bone" or os quadratum.

As regards the limbs of Vertebrates, whilst many differences exist, which will be afterwards noticed, there is a general agreement in the parts of which they are composed. As a rule, each pair of limbs is joined to the trunk by means of a series of bones which also correspond to one another in general structure. The fore-limbs, often called the "pectoral" limbs, are united with the trunk by means of a bony arch, which is called the "pectoral" or "scapular" arch; whilst the hind-limbs are similarly connected with the trunk by means of the "pelvic arch." In giving a general description of the parts which compose the limbs and their supporting arches, it will be best to take the case of a Mammal, and the departures from this type will then be readily recognised.

The pectoral or scapular arch consists usually of three bones, the "scapula" or shoulder-blade, the "coracoid," and the "clavicle" or collar-bone; but in the great majority of the Mammals, the coracoid is anchylosed with the scapula, of which it forms a mere process. The scapula or shoulder-blade is usually placed outside the ribs, and it forms, either alone or in conjunction with the coracoidal element of the shoulder-girdle, the cavity with which the upper arm is articulated. The coracoid, though rarely existing as a distinct bone in the Mammals, plays a very important part in other Vertebrates, as we shall see hereafter. The clavicles are often wanting or rudimentary, and they are the least essential elements of the scapular arch. The fore-limb proper consists, firstly, of a single bone which forms the upper arm (or "brachium"), and which is known as the humerus (h). This articulates above with the shoulder-girdle, and is followed below by the fore-arm (or "antibrachium"), which consists of two bones called the radius and ulna. Of these the radius is chiefly concerned with carrying the hand (or "manus").. The radius and ulna are followed by the bones of the wrist, which are usually composed of several bones, and constitute what is called the carpus. These support the bones of the root of the hand, which vary in number, but are always more or less cylindrical in shape. They constitute what is called the metacarpus. The bones of the metacarpus carry the digits, which also vary in number, but are composed each of from two to three cylindrical bones, which are known as the phalanges (p). Homologous parts are, as a rule, readibly recognisable in the hind-limb. The pelvic arch, by which the hind-limb is united with the trunk, consists of three pieces - the ilium, ischium, and pubes - which are usually anchylosed together, and form conjointly what is known as the innominate done. In most Mammals, the two innominate bones unite in front by ligamentous or cartilaginous union, and they constitute, with the sacrum, what is known as the pelvis.

The hind-limb proper consists of the following parts:

- The thigh — bone or femur, corresponding with the humerus in the fore-limb.
- The bones of the shank (or "cms"), corresponding with the radius and ulna of the fore-limb, and known as the tibia and fibula. Of these, the tibia is mainly or altogether concerned in carrying the foot (or "pes"), and it is thus shown to correspond to the radius, whilst the fibula corresponds to the ulna.
- The small bones of the ankle, known as the tarsus, and varying in number in different cases.
- A variable number of cylindrical bones (normally five), which are called the metatarsus, and which correspond to the metacarpus.
- Lastly, the metatarsus carries the digits, which consist of from two to three small bones or phalanges, as in the fore-limb.

The digestive system of Vertebrates will be spoken of at greater length hereafter; but a brief sketch may be given here of the general phenomena of digestion. All Vertebrate animals are provided with a mouth for the reception of food, and in the great majority of cases the mouth is furnished with teeth, which are used sometimes merely to hold the prey, but more commonly to cut and bruise the food, and thus render it capable of digestion. The food is also generally subjected in the mouth to the action of "salivary" glands, the secretion of which serves not only to moisten the food, and thus mechanically assist deglutition, but also to render soluble the starchy elements of the food.

The food is next swallowed, or, in other words, is transferred from the mouth to the stomach, this being effected by a complicated arrangement of muscles, whereby the food is forced down the gullet (oesophagus) to the proper digestive cavity or stomach. In the stomach the food is subjected to two sets of actions; it is mechanically triturated and ground down by the constant contractions of the muscular walls of the stomach; and it is subjected to the chemical action of a special fluid secreted by the stomach, and called the "gastric juice." This fluid has the power of reducing albuminoid substances to a soluble form, and by its action the food is ultimately reduced to a thick acid fluid, called the "chyme."

Leaving the stomach by its lower aperture (the pylorus), the chyme passes into the intestine, the first portion of which is divided into several sections, but is collectively known as the "small intestine." Here the chyme is subjected to the action of three other digestive fluids; the bile, secreted by a special organ, the liver; the pancreatic juice, secreted by another gland, the pancreas; and the intestinal juice, secreted by certain glands situated in the mucous membrane of the intestine itself. The result of the whole process is that the "chyme" is ultimately converted into a white, alkaline, milky fluid, which is called "chyle." The indigestible portions of the food pass from the small intestine into a tube of larger dimensions, called the "large intestine." Such portions of the food as are still soluble, and capable of being employed in nutrition, are here taken up into the blood, the useless remainder being ultimately expelled by an anal aperture. The last portion of the large intestine is usually less convoluted than the rest, and is called the "rectum."

The fluid and originally soluble portions of the food, and the chyle which is formed in the process of digestion, are taken into the blood, the losses of which they serve to repair. Part of the nutritive materials of the food is taken up directly by the blood-vessels, and is conveyed by the "vena portae" to the liver, whence it ultimately reaches the great veins which go to the heart. The greater part, however, of the liquefied food, constituting the chyle, is taken up, not by the blood-vessels, but by a special set of tubes, which form a network in the walls of the intestine, and are known as the "lacteals." In these vessels, and in certain glands which are developed upon them, the chyle undergoes still further elaboration, and is made more similar in composition to the blood itself. All the lacteal vessels ultimately unite into one or more large vessels which open into one of the veins, so that all the chyle is thus finally added to the mass of the circulating blood.

The blood, then, or nutrient fluid from which the tissues are built up, is formed in this way out of the materials which are taken into the alimentary canal as food. In all the Vertebrata, with the single exception of the Lancelet (Amphioxus), the blood is of a red colour when viewed in mass. This is due to the presence in it of an incredible number of microscopical bodies, which are known as the "blood-corpuscles," the fluid in which these float being itself colourless.

In all the Vertebrata the blood is distributed through the body by means of a system of closed tubes, which constitute the "blood-vessels;" and in all except the Lancelet, the means of propulsion are derived from a Contractile muscular cavity or "heart," furnished with valvular apertures. In the most complete form of circulation, as seen in Birds and Mammals, the heart is essentially a double organ, composed of two halves, each of which consists of two cavities, an auricle and a ventricle.

The right side of the heart is wholly concerned with the "lesser" or pulmonary circulation, whilst the left side is concerned with driving the blood to all parts of the body (systemic circulation). The modifications of the circulatory process will be noticed in speaking of the different classes of Vertebrates, but a brief sketch may be given here of the circulation, in its most complete form, as in a Mammal. In such a case, the venous or impure blood, which has circulated through the body and has parted with its oxygen, is returned by the great veins to the right auricle. From the right auricle a) the blood passes by a valvular aperture into the right ventricle (v), whence it is driven through the pulmonary artery to the lungs. The right side of the heart is therefore wholly respiratory in its function. Having been submitted to the action of the lungs, and having given off carbonic acid and taken up oxygen, the blood now becomes arterial, and is returned by the pulmonary veins to the left auricle (a'), From the left auricle the aerated blood passes through a valvular aperture into the left ventricle (v'), whence it is propelled to all parts of the body by means of a great systemic vessel, the "aorta." The left side of the heart is therefore wholly occupied in carrying out the "greater" or systemic circulation.

MORPHOLOGY

Domenico Cotugno gave the first clear account of the water which filled the space between the dura and spinal cord, and it seems likely that previous descriptions of vapours and spirits around the brain were based on less careful dissections which permitted the Cerebrospinal Fluid(CSF) to drain away and be replaced by air or the gases of putrefaction. The recognition of the CSF as a system must be attributed to Francois Magendie who contributed a series of studies extending from 1825 to 1842.

He noted the continuity of the intracranial with the spinal fluid, and of the fluid around the brain with the fluid within the *ventricles* through a *midline foramen in the roof of the fourth ventricle,* which is commonly called by his name. He considered that the external fluid was formed on the surface of the brain and the intraventricular fluid was formed by the vascular frond-like structures within the ventricles, known as the *choroid plexuses*. He postulated that there was a free exchange between the internal and external fluid by an ebb and flow through his foramen, and that the function of the CSF was to protect the brain and spinal cord and to float them within their bony container.

Magendie's observations appear to be based upon human subjects, as do those of Luschka describing the lateral foramina of the fourth ventricle, and those of Key and Retzius describing the relationship of the subarachnoid space to the arachnoid granulations. However, these observations do not appear in the monographs on vertebrate morphology by Owen and Huxley in the latter half of the nineteenth century; their accounts lack detail of the structure of the fourth ventricle roof and the meninges, and do not mention the CSF as a system.

In this same period following the impact of Darwinism, and doubtless due to the continued emphasis on the importance of the role of the central nervous system in the evolution of man, works devoted to its morphology and evolution appeared but these do not include separate or detailed consideration of a CSF system and the few communications on related structures are limited to consideration of one component, and few species of animals. Nevertheless, the reports that teleost fish have no arachnoid and that the midline foramen (or *metapore*) in the roof of the fourth ventricle can be found in man and several Old World monkeys, but not in the cat might have been expected to arouse the interest of the comparative morphologist had the CSF system been recognised as such, and its evolution considered separately.

The study of any system consisting of fluid in and around a structure such as the neuraxis and contained by fine membranes, presents considerable technical difficulty when the classical method of serial section histology is used, and this may be one reason for its continued neglect. A second reason almost certainly is the concept of a CSF circulation which developed during the first two decades of the twentieth century from the observations of Weed using potassium ferricyanide injections, and those of Dandy and Blackfan upon the production of hydrocephalus. When Cushing gave the Cameron Lectures in 1925 he found enough evidence from the work of his colleagues and that of his own on surgical cases to put forward the concept of a 'cerebrospinal circulatory sector' or 'third circulation' consisting of: the secretion of fluid from the choroid plexuses, its passage through the ventricular system and out of the foramina in the roof of the fourth ventricle, through the basal cisterns and the subarachnoid space to its site of absorption at the arachnoid granulations of the superior sagittal sinus. This concept has become classical and has dominated the understanding of the CSF ever since, although it has never fully satisfied some clinical observations in humans, and many observations derived from experimental work. Furthermore, this concept of the CSF system does not account for its behaviour in animals which lack many of the structural features found in mammals.

In bibliographic terms de Beer can be considered as providing direct continuity with Darwinism in the disciplines of morphology and embryology, the first edition of his book on vertebrate zoology having been supplied with an introduction by T Huxley's son, the zoologist JS Huxley. It contains descriptions of some aspects of the CSF system in animals from amphioxus to the rabbit (in which he describes the arachnoid mater and a foramen of Magendie) but not in full detail for all of the animals studied, nor in the context of a system as such, and the account of the CSF-related structures remained the same in the second edition some forty years later but his further work on embryology elucidated its position in relation to phylogeny and the so-termed 'Law of Re-capitulation'.

Among other twentieth century authorities on zoology Romer in his first and subsequent editions describes the meninges of fishes as a single layer of compact tissue, and states that all tetrapods have an outer dura mater, and an inner layer derived from the neural crest, the latter being divided into arachnoid and pia in mammals; he describes the choroid plexuses of the third and fourth ventricles, but makes no mention of ventricular foramina or CSF flow. Weichert, in his first edition describes the cyclostomes and fishes as having a single-layered meninx (meninx primitiva) with peri-meningeal spaces, and urodele amphibians as having two more-or-less distinct layers - the inner pia-arachnoid and outer dura mater. He describes the three layers in mammmals, and the CSF as passing through three openings in the roof of the fourth ventricle; he does not account for the CSF system in the other animal phyla, and his descriptions remain unchanged in four subsequent editions. Eaton's book on the comparative anatomy of vertbrates describes fishes as having a single meninx, and amhibians, birds, and reptiles as having an outer dura mater, and an inner pia mater close to the brain; the third meninx, the arachnoid, appears in mammals, enclosing the CSF in the subarachnoid space.

Kent, also writing on comparative anatomy, gives no comparisons of CSF structures, but describes a 'sluggish circulation' out of the fourth ventricles to sites of absorption along the roots of the spinal nerves to the lymphatic system, as well as into the venous system via the arachnoid granulations. In his earliest edition he states that the CSF system functions as a protection to the brain, as well as a means of exchanging metabolites and neurotransmitters, while in the third edition he goes so far as to say 'the fluid assists in protecting the central nervous system from concussion'.

Montagna describes the Cushing classical CSF pathway, and notes the limited meningeal differentiation in fishes, and the authoritative account of the vertebrates by Young emphasizes the study of behaviour as well as structure in the evolution of species, and culminates in his classical account of the evolution of the nervous system, but in his detailed presentation, including the phylogeny of the endo- and peri-lymph components of the ear, there is almost no attention to the CSF.

In the last three decades zoologists have reflected interest in the CSF system as a means of maintaining the levels of extracellular metabolites, and have related this to species differences in the ventricular system and other ependymal-derived structures. Kappers extended his earlier interest in the comparative morphology of CSF-related structures to that of the ependymal derivatives also, and included these in a comprehensive text on the comparative anatomy of the brain ; a similar emphasis is contained more succinctly in the small volume by Sarnat and Netsky, the second edition of which contains the most interesting observations upon the activities of the wood-pecker and the particular CSF structure of birds.

Following Cushing's presentation of the CSF system as the 'Third Circulation' it became a topic for monographs presenting its physiology and the laboratory analysis of the fluid as a means of diagnosing diseases of the central nervous system but the observations and experiments contained in these accounts are derived from humans or other mammals - the comparative morhology of the system is poorly represented. The slim volume by Millen and Woollam remains a classical and concise account of the anatomy of the cerebrospinal fluid in the human, but it contains no account of its comparative morphology nor its multiple functions.

Of these functions it is undoubtedly the role of the CSF in the homeostasis of the extra-cellular environment of the central nervous system which has predominated in the basic scientific studies of the last twenty-five years. Among these the notable contributions of Davson to the study of the CSF as a system, and his collation of those of other experts in the field have been presented, with co-authors, in a number of editions showing a varying emphasis around the main theme of homeostasis of the central nervous system.

The relationship of the CSF to the extracellular fluid and the blood brain barriers has remained central in his accounts, but intracranial pressure and pulasatile flow of the CSF is quite fully presented in the last edition and the ontogenetic aspects of the CSF system, including some of the original studies to be presented here. The devoted to the comparative physiology of the CSF, also includes many observations derived from, and related to some of the crucial original studies upon the establishment of communication between the ventricular system and the subarchnoid space in amphibians to be detailed here. The context of these, by inference, is the phylogeny of the CSF flow pattern, but this is not analysed in relation to any protective function of the CSF.

There appears to be a place for the presentation of a full study of the comparative morphology of the CSF system in order to provide not only the structural context that will ensure dynamic studies are species-related, but also as a basis for understanding more of the functions of this system in the support of the evolving brain.

Comparative Morphology as a Method of Study

In a study of comparative morphology it is desirable to examine not only representative species from each class of vertebrates, but also representatives of different orders and even different genera in order to establish the overall pattern of similarities and differences.

Futhermore, while the full study of the morphology of the CSF system in one specimen of any particular species is probably of significance for that species, it does not exclude the possibility of individual variations and pathological abnormalities; observations based upon five or ten specimens could well have an increased value. For example, the experience of examining large numbers of specimens within the class Amphibia revealed important differences between those of the orders Anura and Urodela, but the morphological features in general were the same within each order, and showed only minute species variations.

Most studies of the structure and function of the CSF system hitherto have been made in mammals and, generally, the differences between species is disregarded; the studies which form the basis of this presentation are concerned with the vertebrate kingdom as a whole and the relatively major structural differences which have been found are probably of significance in the evolution of the CSF system.

The structures of the system in the human adult have been well described in modern classical texts already referred to.

Embryological Studies

The morphological studies on the amphibians were of such interest and apparent significance that they were extended to the tadpole and through metamorphosis. Some original studies of the CSF system in Man included in this work are also of its embryological development.

Techniques Used

Fixation All of the specimens were obtained live and the majority were anaesthetised by immersion in a solution containing MS222(Tricaine methane sulphonate, Sandoz) but the birds and mammals were anaesthetised with a calculated dosage of Nembutal. Fixation of the majority of specimens was obtained by intracardiac perfusion with initially Ringer's solution, isotonic for the particular species, followed by dilute Karnovsky solution.

The amphioxus species and the series of frog tadpoles were fixed by immersion in Karnovsky solution. In some species, when initial examination showed fixation to be unsatisfactory, the specimens were decapitated, part of the skull vault excised and the whole immersed in Bouin's solution for 24 hours. In a small number of lampreys and amphibians the ventricular system was injected with either Dextran Blue or the pigment 'Aquadag' in gelatin in order to demonstate the CSF spaces. Injections of the ventricular system with Batson's solution were performed in fresh-frozen dogfish specimens and anaesthetised amphibians; after maceration with concentrated, boiling potassium hydroxide, ventricular casts were obtained.

- *Decalcification* Specimens were immersed in solutions of RDO for 24-48 hours for decalcification.
- *Clearing* The lamprey and majority of the amphibians were cleared in cedar wood oil, but all of the remaining specimens were cleared in terpineol.
- *Tissue processing* On the first day 50%,60%,and 70% alcohols were used during the day, and 80% alcohol overnight. On the second day 90% and 95% alcohol followed by three changes of 100% alcohol, alcohol and cedar wood oil in a 1:1 mixture for one hour, and pure cedar wood oil overnight. On the third day cedar wood oil and toluene, 1:1 for one hour followed by pure toluene for one hour.
- *Embedding* The majority of specimens were embedded in wax using three changes and then left overnight in the fourth change of warm wax. They were finally evacuated in fresh warm wax to remove all air bubbles and then embedded in paraffin wax or paraplast.

- *Section cutting* Serial sections were cut in a transverse plane at a thickness of 6um for smaller specimens but in specimens which were large and easily fragmented the sections were cut at 8um. A few specimens were sectioned longitudinally.
- *Staining* The specimens were mounted on slides serially, and stained in rotation using the standard Haematoxylin and Eosin stains for general orientation and identification of cell types, Van Gieson's stain to demonstrate collagen in the membranes surrounding the brain and in the connective tissue within the neuraxis; the third slide in the rotation was stained with Masson's trichrome which shows collagen, reticulin, amyloid and mucin as green, nuclei as black and muscle, red blood cells and most cytoplasmic granules as red, or Mallory's trichrome with which the collagen, reticulin and basophilic substances are a deep blue and other structures, by contrast, light red or orange, and the nuclei stain purple. In a few series the trichrome stain was Martius scarlet-blue which stains connective tissue blue, fibrin red, and nuclei purple to black.
- *Mounting* The slides were mounted using Histamount.

Animals Chosen and Preparations

The amphioxus proved initially difficult to fix and section without shattering, and the most satisfactory specimens were obtained after immersion in sea water, followed by immersion in Karnowsky solution over 24 hours. Full serial sections satisfactory for study were obtained from four specimens.

The lamprey specimens were satisfactorily fixed by intracardiac perfusion with Karnovsky solution and serial sections were taken of the body and tail sections in some specimens. Six very satisfactory series were obtained, including one sectioned longitudinally. The one dogfish obtained in the live state was well fixed with intracardiac perfusion of Karnovsky solution followed by immersion of the rather large head after the cranial vault was removed. The sections were cut at 8um.

Seven goldfish specimens were prepared suitable for study, including one which was longitudinally sectioned. Two of the lungfish were prepared and one of these, sectioned at 10um, produced a good series. Over 100 specimens of *Rana temporaria* were prepared, including a special series of 10 adults sectioned for detailed study of the roof of the hindbrain, and two series of tadpoles at various stages of development up to matamorphosis. Specimens of the following Anurans were also studied: *Bufo bufo, Xenopus, Rana pipiens, Hyla arborea, Rana catasbiana,* and *Rana ridibunda.*

Five adult salamanders were prepared and one of these was longitudinally sectioned. The tail was studied in some specimens. In addition the following Urodeles were studied: *Triturus vulgaris* and *Necturus maculosa.*

Two turtles and nine lizards were used in the study. In preparing the turtles, post-fixation of the excised head in Bouin's solution was performed, and after decalcification the specimens were softened in phenol for two days.

A similar modification of technique was used for the chick and pigeon; four birds were fully studied by serial sections and an additional pigeon was prepared and the various membranous layers at the cerebello-medullary junction were dissected and examined. The mouse also had post-fixation in Bouin's solution after decapitation, removal of the cranial vault, and preparation of the brain by an initial longitudinal section between the cerebral hemispheres, followed by transverse sections from the midbrain caudad.

CHEMICAL CONSTITUENTS OF LIVING ORGANISMS

More than is often realised, scientific progress is dependent on the coining of apt phrases and catchwords. The study of granules in cells was certainly fostered by Benda's introduction of the word "mitochondria." So too the introduction of the word "vitamin" was a factor in the development of a field which had never prospered as long as vitamins were called by such awkward terms as "accessory articles of the diet."

In the study of living material the concept of a cell theory or a cell doctrine has attracted the attention of students and workers to the fact that even the most diverse types of animals and plants are constituted of the same type of unit. It is probably true that living material is as basically similar in chemical composition as in microscopical appearance; and yet this important fact, unheralded by any catchword, has never received the emphasis it deserves. True, the concept of protoplasm has had a wide influence, but the term "protoplasm" is commonly used more in a morphological sense than in a chemical one.

At the present time, indeed, one is not quite sure as to just how constant protoplasm is in a chemical sense. Numerous analyses have been made of the organs, tissues, and cells of living organisms, but these studies have been made by biochemists who have generally had little or no interest in the protoplasm concept, and were not concerned with the degree of uniformity shown by diverse types of living material.

Just as the cytologist has attempted to explain life by a study of the minute details of microscopic anatomy of cells, so the chemicallyminded biologist might seek to explain life by studying the chemical composition of living substance. As a matter of fact, however, students of the chemistry of living material have never been very optimistic in this direction. Obviously, almost all types of chemical analysis first lead to death of the cell, so that with ordinary chemical methods one can only obtain information regarding that which *was* once alive. Indeed, the chemical properties of living protoplasm are but little understood. We do not clearly understand what it is that distinguishes living protoplasm from dead, although it is certain that chemical changes do accompany and follow the death process.

Chemically, it is possible to isolate from living material various types of organic compounds. Usually, it is fair to assume that the compounds isolated were present as such when the protoplasm was alive. And yet there is good reason to believe that frequently substances obtained from cells were really in life parts of more complex compounds. So, for example, a carbohydrate extracted from living tissue may before the extraction have been bound to a protein; similarly fats or fat-like substances gotten from cells may in the living cells have been tied to protein.

Also, individual proteins which are obtained in extracts of tissues may originally have been combined with other proteins. This is now well known to be true in the case of muscle. Thus it appears that all the organic compounds known to constitute the mass of the living protoplasm may to a large extent be tied together. Also they may be tied to inorganic constituents such as phosphoric acid, sulfate, metals, alkaline earth cations or even water.

One of the most interesting aspects of the chemistry of living material is the ability of the chemist to extract from it certain catalytic substances or enzymes. These enzymes can be isolated in rather pure form and their effect on chemical reactions can be followed in test tubes. But even in the case of these enzymes, it is not certain that the extracted enzymes are in the same state as they were in the cell. Thus von Euler and von Euler urge that within the cell enzymes may exist in the form of complexes. Following extraction, the complexes are broken up into simpler constituents.

If so many substances of a cell are tied together, we might be inclined to think that protoplasm consisted in reality of giant molecules containing practically everything of importance to the life of the cell. Such ideas have been held in the past. For example, Verwor n conceived of a giant protoplasmic molecule, which he called biogen.

Verworn's ideas of the nature of biogen could hardly be held now, and indeed the modern biochemist prefers to think of protoplasm as composed of many organic compounds existing separately within the cell. If, however, at some future time viruses come to be regarded as truly living, we might have to extend our idea of protoplasm so as to include materials consisting of just one type of chemical compound.

ELEMENTS IN PROTOPLASM

There may be uncertainty as to exactly which compounds exist within the living protoplasm, or as to how the various compounds are tied together, but it is a relatively simple matter to determine exactly which elements are present and in what amount. The ordinary methods that the chemist uses for qualitative and quantitative analysis can be employed. In addition, spectographic methods are very helpful, and they can be used for minute amounts of material.

In the newer spectrography, quantitative determination of elements has been made possible by the use of photoelectric devices. The radiation emitted by an ignited substance is split up into its constituent wave lengths and the spectrum may then be recorded on photographic film or plate. Then the intensities of the spectral lines are determined by means of a densitometer or microphotometer and these intensities are compared with a calibration curve obtained from known substances in known amounts. Or without photographic film one can compare the intensity of spectral emission lines with the intensity of the lines given by known quantities of known elements.

The flame photometer is now widely used in physiological laboratories and is especially suitable for the rapid determination of potassium and sodium. In this apparatus, a solution containing sodium, potassium or other elements is fed into a flame and the emission spectrum is then compared with that of a solution containing a known amount of the element. However, it must not be supposed that the determination of elements within living material is always easy. There are difficulties which are often not fully realised. True, standard methods of quantitative analysis are applicable, but such methods can be properly used only if the material to be examined is in true solution. Usually, moreover, inorganic elements to be recognized must be in ionic form.

It is not always an easy matter to dissolve living substance completely and to free it from carbon and from other substances which would interfere with the reactions of quantitative analysis. In order to proceed with such an analysis, the living tissue must first be ashed. Because of the fact that elements like chlorine and potassium volatilize at not too high temperatures, most biochemists use what are called wet-ashing procedures. The tissue is heated with concentrated nitric or sulfuric acid or both, and in this way the carbon it contains is oxidized.

Sometimes to complete the oxidation, after ordinary acid treatment the material is dried and treated with perchloric acid. Eventually it should dissolve completely but sometimes it does not. Or even if it does, the element being tested for may still be combined with organic materials and as a result improper values are obtained when standard methods of analysis are applied. Thus, for example, when muscle is tested for chloride by the commonly used Van Slyke method, the muscle is dissolved by heating it in concentrated nitric acid, and at the same time the chloride is precipitated by silver nitrate.

Titration for chloride is then done by the standard Volhard method. Actually, this procedure, though it has been widely used, gives incorrect values for chloride. If the muscle tissue is really ashed by ignition in a tightly sealed container such as a Parr bomb, values for chloride can be obtained which are approximately twice those that can be obtained by the Van Slyke method. Even with the spectrographic methods, difficulties in ashing may be encountered. In the flame photometer the material is used in solution. A small sample of the solution is then ignited.

But in order to obtain a representative sample, it is necessary to have either a clear solution or an even suspension. Sometimes in preparing material for analysis, a few relatively large flakes of precipitate may appear. This can be very annoying, and can prevent proper analysis. Similar difficulties may arise in ordinary spectrography. Thus, in interpreting the results reported in the literature, it is often necessary to view these cautiously. In the discussion that follows, no attempt will be made at completeness.

Over 34 elements are commonly found in plant and animal tissues. Of these, carbon, hydrogen, oxygen and nitrogen are constituents of such organic compounds as proteins, fats and carbohydrates; also, to a less extent, sulfur and phosphorus.

Other elements may likewise be combined with the proteins of the cell, but ordinarily they do not form an intimate part of the protein molecule. In addition to the elements already mentioned, the following are among the elements found in practically all types of living substance: potassium, sodium, calcium, magnesium and chlorine. However, many plants can grow in the apparent absence of sodium and, indeed, some of the common nutrient solutions used for the growth of plants contain no sodium.

Lithium is typically present in marine animals, but according to Fox and Ramage, lithium, although widespread, is not a universal constituent of protoplasm. Rubidium is also rather widespread in its occurrence. It was found in all human tissues examined except bone, being especially abundant in heart and striated muscle. Although higher plants seem to require calcium, some fungi are said to grow normally without it. It is by no means an easy matter to remove traces of sodium or calcium from nutrient solutions, so that it is difficult to determine whether or not living material can exist in the total absence of these elements.

Strontium is close to calcium chemically and appears to be able to replace it in some biological systems. A small amount of strontium is generally present in marine invertebrates . Magnesium is a constituent of chlorophyll, and is therefore found in all green plants. Although sea water itself contains several times as much magnesium as calcium, the animals which live in sea water commonly contain less magnesium than calcium. This is not always true, however, and some gastropods including the very common marine snail Littorina littorea contain more magnesium than calcium. Some insects are also richer in magnesium than in calcium. Rats cannot live without a small amount of magnesium.

If deprived of it, the calcium-magnesium balance is upset, and convulsions and death follow. Boron appears to be an essential constituent of citrus plants, tobacco, and tomato. It is also found in small amounts in Marine Invertebrates. Aluminum is a very common constituent of living material, especially in plants. It is also found in animal tissues. Heavy metals, although not present in large amounts, are very important for life. In organic combination they act as carriers of oxygen (in the blood) and as oxidizing catalysts within cells. Especially important are iron and copper.

Iron is a necessary constituent of the blood cells of all vertebrates and of many invertebrates. As Warburg first showed, cellular respiration is primarily dependent on the presence of iron-containing catalysts believed to be present in all living cells.

The blood cells of many invertebrates contain copper and the blue colour of crab and limulus blood is due to the presence of this element. Lindow, Elvehjem, and Peterson have shown a widespread distribution of copper in living tissues; Sheldon and Ramage conclude that it is a universal constituent of living protoplasm.

This opinion is shared by Marston. The literature on the copper content of living systems is now very large, mainly because of the importance of copper in cell metabolism. Zinc is widely distributed among both plants and animals. Galtsoff found it in many marine invertebrates. It has also been reported for the silkworm and for vertebrates generally. Gettler and Bastian describe a simple method of detecting zinc in tissues and they find rather high amounts (50 mg. per kilo) in human liver and kidney.

Keilin and Mann showed that zinc is a constituent of the important enzyme carbonic anhydrase and Smirnov finds that the zinc content of the red blood cells of various vertebrates runs parallel to their carbonic anhydrase content. Zinc is also present in blood plasma.

The retina, iris, and choroid of vertebrate eyes and especially the eyes of fish are rich in zinc and this is also to be correlated with a high carbonic anhydrase content of these tissues. In view of the fact that the insulin molecule is thought to contain zinc, it is not surprising that the pancreas should be rich in this element. Manganese is also an important element.

It is essential for the growth of rats and chickens, and also for the growth of various types of plants. Webb found manganese generally present in invertebrates; according to Vinogradoff, ants have an especially high content. Human blood contains only a small amount of it. In almost all plant and animal tissues, it is possible to detect cobalt and nickel by means of extremely delicate tests. The quantities found range from 0.002 to 2 mg. per kilo.

All types of animals that have been tested contain molybdenum; tunicates are especially rich in it. Marston (loc. cit). believes that molybdenum is probably always present in living matter. Some metallic elements are of rather sporadic occurrence. Small amounts of arsenic and of silicon have been found in various tissues of man and higher animals. Ohlmeyer and Olpp found quite appreciable amounts of silicon in the blood, lung and pancreas of the calf. It had previously been found in connective tissue, hair and feathers.

Some plants seem to have a definite need for silicon. Webb found tin in the starfish, the sea urchin, and in a gastropod mollusk. There is a literature on the presence of vanadium in the blood cells of ascidians. The element is also present in other tissues of ascidians but not in detectible amounts in rat tissues. Vanadium occurs in a wide variety of plants. Traces of uranium have been found in the fungi Aspergillus niger and Ustilago carbo; also in human kidney, brain, thyroid and blood.

The rare element gallium is apparently essential for Aspergillus. Many human organs contain mercury in small amounts. In some parts of the world gold is found in the soil and is taken up by plants. Deer eating such plants accumulate gold in their horns, hoofs and hair. There is a large body of evidence to show that metallic elements present in traces have a profound effect on the life of various types of plants. The subject is of great practical importance. Of the non-metallic elements, as already mentioned, chlorine is probably found in all types of living material. For many years, it was believed that chlorine although present in tissues generally, did not occur within the cells but only in the spaces between the cells.

This doctrine was at one time held for various types of vertebrate tissues and organs, but especially for muscle.

Analysis of muscle by the standard Van Slyke method showed relatively low concentrations of chlorine, and this could all be accounted for if it were assumed that the tissue spaces of the muscle contained chlorine in the same concentration as that in which it occurs in the blood.

As a matter of fact, as already noted, the Van Slyke method gives improper values for the chloride of muscle. Moreover, chloride in reasonably high concentrations can actually be demonstrated in isolated frog muscle fibers.

Fluorine is of little importance for animals and plants, for with few exceptions it occurs only in minute amounts and it is not known to have any well-defined function. The same applies to bromine. As is commonly known, iodine is an important constituent of the thyroid gland of vertebrates. Iodine is also abundant in marine brown algae and is obtained commercially from this source.

COMPOUNDS of PROTOPLASM

Although it is simpler to determine the chemical *elements* present in living substances, one can also attempt an analysis of the chemical *compounds* which are present.

It is not difficult to determine the percentage of various classes of substances. Numerous such analyses have been made for tissues, organs, and whole animals.

Analyses of cells are less common. The following table gives the percentage composition of the egg of the sea urchin, Paracentrotus lividus, and is taken from Ephrussi.

	Per cent
Water	77.3
Protein	15.18
Fat and lipoid	4.81
Carbohydrate	1.36
Ash	0.34
	98.99

This may be compared with an analysis of the human body shown in the following table.

	Per cent
Water	66.0
Protein	16.0
Fat and lipoid	13.0
Carbohydrates	0.61
Ash	5.0

In the above analysis the figures for fat and ash are high, doubtless because the body as a whole contains many non-protoplasmic materials. As compared with animals, plants are relatively rich in carbohydrates. Analyses of plant tissues show a content of carbohydrate material which may be equal to or even in excess of the protein content. *Water.*

As already indicated, water is the chief chemical constituent of protoplasm. Concerning it Adolph says: "The vortex of vital activities is, with respect to water, more than analogous to the whirlpool.

Without water both the body and the pool would be but paleontological specimens, showing only (in Huxley's phrase) where activity has been." Most of the determinations of the water content of living materials have been made by drying these materials in an oven at temperatures slightly above the boiling point of water.

The loss of weight is assumed to represent the original water content. This procedure doubtless involves some loss of volatile matter other than water. Also, because of the presence of humidity within the oven all the water is not driven off. Use of a vacuum oven, or of circulating hot air, can reduce the humidity error.

As a matter of fact, the two types of error tend to neutralize each other, although the error due to incomplete dryness in the oven is relatively large. Because of the humidity error, many of the older determinations of the water content of organisms and tissues may be too low. A simple and rapid method of water determination is to place tissues or small organisms in toluene or some similar liquid and distill off the water. This is then collected and its volume determined. There is also a titration method for water determination. Some organisms have very high percentages of water.

Thus according to Vernon, the water content of the ctenophore Cestus is 96.2 per cent; the jellyfish Rhizostoma has 96.0 per cent, the tunicate Salpa 96.3 per cent, and the nudibranch Tethys 95.3 per cent. Paramecium has 79 per cent water.

Even in the tissues of higher animals the water content may be rather high. Thus in the gray matter of the brain and in the kidney the water concentration is above 80 per cent; so that in these organs there is actually a higher concentration of water than in the fluid blood.

The lowest concentrations of water are found in seeds and in certain types of animals which have the power of drying up.

Rotifers and tardigrades normally live in ponds and puddles. In dry seasons, as the ponds or puddles dry up, the rotifers lose practically all their water.

In such dried-up forms protoplasmic activity ceases almost entirely. As a matter of fact, protoplasmic activity is associated with water content. Young and growing animals and plants are richer in water, and are more active than somewhat older organisms.

Most of the water of a living cell is free, that is to say, it is not bound chemically to other chemical constituents of the protoplasm. However, a certain percentage of water may be "bound."

The subject of bound water is a controversial one and at present is in an unsettled state. Proteins can bind water chemically. Certainly this is true of dried proteins or of crystalline proteins. But whether the water bound in a crystal remains bound when the crystal dissolves is of course doubtful. Bound water is defined in various ways.

For most authors bound water is water which is not available for dissolving solutes.

Such water may be bound chemically or it may be held physically either by adsorption or by being trapped in the interstices of a gel or jelly. According to one author, water is free if it evaporates according to the usual kinetics of evaporation, but is bound if it is retained by the medium and evaporates less actively than free water.

Concerning the amount of bound water in colloidal solutions of proteins and other substances, there is dispute. Newton and Gortner determined percentages of bound water in plant juices by studying the effect of adding sugar on the freezing point of these juices.

Chambon and Gouttebarge found large amounts of bound water in plant gums, mucilages, and pectins, as well as in animal materials such as the vitreous humour of the eye or the mucus secreted by the foot of a snail. They believe that there is a correlation between the amount of bound water and the uronic acid content of a substance.

Table. Inorganic Composition of Longitudinal Muscles of Caudina Chilensis (The Figures Represent Equivalents Per 100 Gm. Fresh Weight)

Cations		Anions	
K	0.01385	Cl	0.01220
Na	0.01910	SO4	0.01305
Ca	0.01775	CO3	0.01700
Mg	0.00785		
Total	0.05855	Total	0.04225

Chambon and Gouttebarge used the Nicloux method, in which the amount of water available for the solution of alcohol is determined. In this method very dilute solutions of alcohol are used and the method is available for living systems. Nicloux himself used his method in this way and he determined the amount of bound water in small (whole) fishes and in various tissues of larger fishes and frogs. This work has perhaps not had the attention it deserves.

This analysis differs from most analyses of muscle in that it shows more sodium than potassium. Typically muscle is much richer in potassium than in sodium. It will be noted that the sum of the cation equivalent weights is decidedly greater than that of the anions.

Organic Constituents

Living substance contains three main types of organic constituents. Of these the proteins and carbohydrates are typically water-soluble substances. It is true that some proteins and some carbohydrates are not soluble in water or in dilute salt solutions, but insoluble proteins are relatively rare, and insoluble carbohydrates are easily broken down to soluble compounds.

On the other hand, the third class of substances characteristic of protoplasm is distinguished by a thorough-going insolubility in water. Indeed, all protoplasmic constituents not soluble in water, but soluble in fat solvents, are grouped together as lipids. The term "lipids" can then be taken to include, first, the true fats, secondly the phosphatids which are compounds of fatty acid, phosphoric acid, and organic base, and, thirdly, the sterols.

The latter bear no chemical kinship to the fats, but resemble them in being insoluble in water but soluble in ether. Although not so abundant in living material as are the proteins, the lipids are indispensable for life. Insoluble in water, they are nevertheless finely dispersed in the watery protoplasm.

This is in part due to the fact that the lipids are emulsified; in part, also, to the fact that the lipids may combine with proteins to form water-soluble compounds. The nature of these compounds is not at all well known, but it seems certain, as already noted, that they do occur in protoplasm. The combination of lipid with protein is but one example of the fact that within the living cell various types of chemical compounds may be linked to each other. At the present time, our knowledge of actual combinations of organic substances in the living protoplasm is meager. What we know chiefly is that the organic substances which do constitute the protoplasm are of such a nature that they might well be assumed to combine.

Thus, if some of the proteins contained in cells behave primarily like acids and others like bases, it is only natural to assume that these proteins might be linked together in life. Indeed the existence of combined proteins is well known. Carbohydrates may be combined with proteins not only in true chemical combination but also the carbohydrates may be tied to the proteins by loose secondary linkages.

Various alkaline and alkaline earth cations can be and undoubtedly are bound to the proteins of the protoplasm, and they may also be linked to lipids or even to carbohydrates. Perhaps also anions may be bound to the proteins of cells. Indeed, although no one at the present favours the old biogen concept of protoplasm, there is a growing fund of knowledge to indicate that many of the chemical compounds isolated from dead cells may in the living organism be linked by chemical bonds.

OSMOTIC RELATIONS OF CELLS

In view of the fact that living cells are surrounded by osmotic membranes, and are thus tiny osmotic engines, it is important for the physiologist and for the general biologist to know something about osmotic phenomena and the laws of osmosis.

Indeed, in any investigation in which cells or tissues are immersed in a solution it is essential that this solution have the same osmotic properties as the fluids in which the cells and tissues are normally bathed. Similarly, if a solution is injected into the blood stream of an animal, it should not differ markedly in osmotic properties from the blood itself.

Many a biological investigation has been spoiled by a failure to observe this rule, or by an ignorance of how to arrive at a proper osmotic concentration. In the following discussion no attempt will be made at completeness, and very little attention will be paid to theories of osmosis or to mathematical treatment of osmotic laws. Moreover, although solutions in all types of solvents show osmotic behaviour, we shall confine our attention to watery solutions, for these are the only types of solutions with which we shall have to deal in living systems. When two solutions are separated by a semipermeable membrane, there is a flow of solvent toward the solution of higher molecular concentration.

This constitutes osmotic flow, and the pressure developed as a result of this flow is osmotic pressure. A semipermeable membrane is a membrane which permits solvent to pass, but not dissolved substance or solute.

Actually no perfect semipermeable membrane is known; at any rate, we know of no membrane which permits water to pass and resists the passage of all dissolved substances. However, there are membranes which are permeable to water and impermeable to certain dissolved substances, and if we choose the proper solutions, these membranes are truly semipermeable.

Permeability of the Cell Membrane

The osmotic behaviour of a living cell depends in large measure on the characteristics of the membrane which surrounds it. As pointed out in the previous chapter, only those substances which are incapable of passing through it or which pass through with difficulty produce osmotic pressures against it. It is important, therefore, to have information concerning the permeability of the plasma membranes of cells if we are to have a true picture of the osmotic relationships. There are also other reasons for studying cell permeability: every cell is dependent on its environment.

It obtains nutritive substances from the fluids which surround it, and to these surrounding fluids it passes out excretory substances. All of this exchange takes place through the cell membrane, and the rate of exchange is of course governed by the permeability of the membrane. Moreover, this permeability is not constant but variable: it changes with the physiological state of the cell; it is influenced also by external conditions such as temperature or light or the chemical constitution of the environment.

In the changing permeability of the cell many authors have thought to find an explanation for various types of vital activity.

These authors apparently regard cell permeability as the chief variable of those which are known or knowable in the life of the protoplasm. Moreover, some authorities believe that the differences in electric potential between various parts of living systems and the changes in these potentials during vital activity are primarily due to membrane permeability differences.

Finally, a knowledge of the permeability of the osmotic membrane or plasma membrane of the cell gives information as to the essential properties of this membrane, and may even lead to conclusions as to its chemical make-up. To avoid confusion, the concept of permeability needs to be clearly defined.

According to Brooks and Brooks, it is "the rate of movement of a substance through the permeable layer under a given driving force. This definition involves two concepts which must be kept sharply distinguished: on the one hand, permeability—a property of a membrane or region of a cell: on the other hand, the driving force which may be independent of any property of the membrane."

The driving force is primarily due to a difference in concentration of a substance on the two sides of the membrane or permeable layer. This driving force becomes progressively less as equilibrium is approached, finally reaching zero at equilibrium. "Permeability measures the rate at which equilibrium is approached, that is, what fraction of the total change necessary for the attainment of equilibrium is acomplished in unit time."

In non-living systems, ordinarily when a membrane separates two concentrations of a given substance to which the membrane is permeable, equilibrium is reached when the concentrations at both sides of the membrane are equal. But in a living cell such an equilibrium may never occur and the cell may reach a condition of no further change and yet possess either more or less of the substance than the surrounding medium. This may be due in part to adsorption (at the surface or within the cell) or to chemical reactions. In addition, it must be remembered that the cell is alive and that a portion of its vital energy may be expended in furthering the accumulation or the disposal of one substance or another.

Thus, when there is no further change in the ratio of concentration (of a given substance) inside and outside of the cell, a true equilibrium may not exist, but rather a dynamic equilibrium. Such an equilibrium is generally referred to as a "steady state." Normally a given cell is exposed to water plus a variety of substances in solution.

The membrane of the cell has a definite, although doubtless a somewhat variable, permeability toward each of the substances to which it is exposed, as well as to any other substance which may be added. In the past confusion has arisen because authors have frequently written of permeability in a rather loose fashion.

Thus it is not sufficiently accurate to say that the permeability of a cell increases-we must know to what substance or substances the permeability is increased. Recent literature indicates that permeability to one substance may increase, whereas permeability to another may decrease.

In discussing the permeability of the cell we shall for the most part consider only the plasma membrane. Actually, when a dissolved substance penetrates a plant cell, other membranes than this are also involved. The nonliving cell wall of plant cells is ordinarily permeable both to water and to dissolved substances, but in special cases, as for example in barley and wheat seeds, the cell wall offers resistance to the passage of dissolved substances. Plant physiologists often study the passage of substances into the large vacuole of a plant cell; in order to reach this vacuole from the exterior of the cell, substances must pass through not only the plasma membrane but also a membrane or film which separates the vacuole from the protoplasm.

It is well to remember this in interpreting the many studies in which the penetration of substances into plant vacuoles has been measured. Of course, it is possible that the properties of the membrane lining the vacuole are similar to those of the outer cell surface, but this has never been proved, and there is indeed some evidence to the contrary.

There have been many studies on cell permeability. The earlier investigations were concerned mostly with the qualitative information as to whether or not the cell membrane permitted the passage of this or that substance, or as to whether various conditions and reagents caused increase or decrease of permeability.

In the more recent literature there has been an increasing effort to make quantitative determinations of the speed of entrance of various substances into the cell (or out of it). We shall consider both the permeability of the cell membrane to dissolved substances and to water.

For the older literature on permeability the monograph of Stiles is very useful. More recent books on permeability are those of Brooks and Brooks and Davson and Danielli. The discussion in Höber's Physical Chemistry of Cells and Tissues may profitably be consulted. Wilbrandt has published some useful reviews.

The results obtained in the study of cell permeability are not always in agreement. This is due in part to the fact that cells of different types do not always show the same permeability relations toward dissolved substances. Another reason for disagreement lies in the fact that the various methods are not equally dependable. Indeed, some of the indirect methods are open to serious question. It is important to examine the various methods critically, as Stiles and Jacobs have done.

According to Jacobs, "It will appear that no single method by itself is entirely reliable, and the obvious conclusion to be drawn is that in any given investigation as many and as diverse methods as possible ought to be employed before an attempt is made to base generalizations on the results obtained."

One difficulty common to all methods is the sensitivity of the plasma membrane. When cells are injured the semipermeability of the membrane is apt to be more or less completely destroyed. Care must be taken, therefore, to avoid injury as much as possible.

If, for example, one is studying the permeability to acids, it is hard to be sure that even in dilute solutions the acid does not produce some measure of injury.

Some methods depend on the exposure of the cell to hypotonic or hypertonic solutions, and such solutions may also change the permeability of the membrane.

7

Circulation in Animals

Materials formed in one part of the body have to be taken up to other parts where they are needed or to be got rid of. This is an essential requirement of most animals. This function is performed by the body fluids.

FUNCTIONS OF CIRCULATORY SYSTEM

The functions performed by the circulatory system in all animals are, Transport of nutrients like glucose, amino acids, vitamins, minerals.

Types of Circulatory System

In the closed type of circulatory system, the blood remains inside the blood vessels and does not come out. The blood flows from arteries to veins through small blood vessels called capillaries. The closed type of circulatory system occurs in most of the Annelids, Cephalopods and Vertebrates including man.

Open Circulation System in Cockroach

The heart of the cockroach is elongated, thick, muscular, tubular and 13-chambered. It lies in the pericardial sinus of the haemocoel. Each chamber of the heart receives oxygenated blood from the dorsal sinus through one pair of slit like openings called ostia. The heart contracts in a postero anterior direction and the blood also flows posteroanteriorly. The alary muscles are responsible for the circulation of blood. The first chamber leads into an aorta, which opens in the head sinuses which are connected to the pericardial sinus through perineural and perivisceral sinuses.

Closed Circulatory System in Humans

The organisms have a thick body wall to prevent the evaporation of water, so exchange of materials between the body cells and the environment by diffusion is not possible.

Blood Vessels

These are hollow, tubular vessels which conduct the blood from the heart to the tissues and from the tissues to the heart.

There are 3 type of blood vessels:

- Arteries,
- Capillaries and
- Veins.

Components of Circulatory System

The WBCs are larger in size than the RBCs but much lesser in number. They are different from the erythrocytes in the following aspects.

Heart—Shape and Position

It is a thick muscular, reddish brown, conical organ present in the mediastinal space of thoracic cavity between 2 pleura enclosing the lungs. Its broader side is called base and it is forward and upward while the pointed side called apex is backward and downward. It is 9 cm broad and 12 cm long and about 300 gms in weight.

Heart—Cardiac Cycle

This phase involves the contraction of the 2 auricles, pushing the blood into the respective ventricles. There is no back flow of blood due to the presence of the bicuspid and the tricuspid valves. The atrial systole takes 0.1 second.

This is followed by the atrial diastole when both the auricles relax simultaneously. This is about 0.7 seconds.

Circulation of Blood through the Mammalian Heart

The aorta or the great artery arising from the left ventricle takes the oxygenated blood to the body organs through a number of arteries. From these organs, the deoxygenated blood is collected by the superior and inferior vena cava and brought back to the right auricle. The right auricle pumps the deoxygenated blood into the right ventricle.

Nervous Regulation

The heart is regulated to a large extent through the central nervous system. The heart nerve branches from the vagus nerve from the medulla oblongata and the sympathetic nerve fibres from the spinal cord.

Rate of Heart Beat

Pulse is defined as the wave of distension and recoiling felt in the radial artery due to the contraction of the left ventricle which force about 70-90 ml of blood into the already full aorta.

Causes of Blood Pressure

The pressure existing in the arteries is called arterial blood pressure. The blood pressure is high during systole and is called systolic blood pressure. It is low during diastole and it is called diastolic blood pressure. The difference between these two pressures is called pulse pressure.

Blood Related Disorders

Persistently having a resting blood pressure of more than 120 / 80 mm Hg is called hypertension. In such a condition, the heart has to work harder to pump the required amount of blood to the various organs through narrowed arteries.

ECG - Electrocardiography

ECG is a graphic display or recording of the electrical variations produced by heart muscle during a cardiac cycle. The activity of the heart create an electrical field which is conducted through the surrounding body tissues to the surface of the body.

Components of Human Lymphatic System

The lymphatic system is an accessory circulatory system which transports lymph, a fluid similar to plasma from the intercellular spaces of tissues to the blood. It is a one way route for the movement of interstitial fluid to blood. The lymphatic system can carry proteins and large particle matter from the tissue spaces into the blood.

Immunity and Immune System

Animals encounter many potentially dangerous microbes in air, water and food. So they have involved defence mechanisms against the disease causing germs Immunity is a defence mechanism without which we will fall a prey to all parasitic microorganisms.

Internal Defence

Second line of defence or body's internal defence is carried out by leucocytes, macrophages, inflammatory reactions, fever, interferons and natural killer cells. These operate together to check the damage to the body by the pathogens.

Specific Defence Mechanism

This mechanism provides protection against specific foreign materials and is often called the immune system. This system forms the third line of body's defence against microbes and harmful molecules. The most important characteristic of the immune system is that its cells have an ability to recognise body's own cells and macromolecules, from those which are foreign invaders or non-self.

Cells of the Immune System

Lymphocytes are the main cells of the immune system. They arise from stem cells present in the liver in the foetus and in the bone marrow in the adult.

Humoral Immune System

The plasma membrane of each B cell should be sensitised by contact with a specific antigen for the release of antibodies. The plasma cells do not migrate to the site of infection but act through the lymph. So they form the humoral immune system. The B lymphocytes are short lived and are replaced every few days from the bone marrow.

Cell Mediated Immune System

The cellular immune response is given by T - cells. There are separate T - cells for each type of antigen that invades the body. The life span of the T - cells is 4-5 years or even longer. There are four types of T - cells.

Immunity

Immunity is the ability of an organism to recognise the foreign material or chemicals that enter the body and to mobilise the cells and cell products to quickly remove the foreign material.

Types of Immunity

The type of immunity inherited by the organism from the parents and protects it from birth throughout life is known as innate or inborn immunity.

THE SQUID HEARTS

This group of marine invertebrates has solved the problem by having separate pumps:

- Two gill hearts to force blood under pressure to the gills and
- A systemic heart to force blood under pressure to the rest of the body.

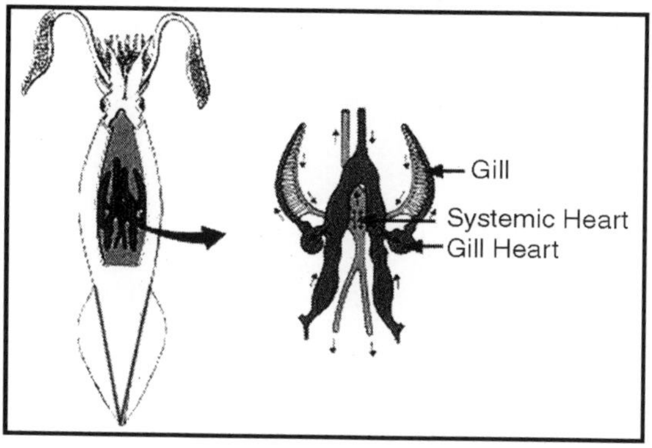

Fig. : Squid Hearts

THE FROG AND LIZARAL HEART

The frog heart

The frog heart has 3 chambers: two atria and a single ventricle.

- The atrium receives deoxygenated blood from the blood vessels (veins) that drain the various organs of the body.
- The left atrium receives oxygenated blood from the lungs and skin (which also serves as a gas exchange organ in most amphibians).
- Both atria empty into the single ventricle.
- While this might appear to waste the opportunity to keep oxygenated and deoxygenated bloods separate, the ventricle is divided into narrow chambers that reduce the mixing of the two blood.
- So when the ventricle contracts,
  - Oxygenated blood from the left atrium is sent, relatively pure, into the carotid arteries taking blood to the head (and brain);
  - Deoxygenated blood from the right atrium is sent, relatively pure, to the pulmocutaneous arteries taking blood to the skin and lungs where fresh oxygen can be picked up.
  - Only the blood passing into the aortic arches has been thoroughly mixed, but even so it contains enough oxygen to supply the needs of the rest of the body.
- Note, that in contrast to the fish, both the gas exchange organs and the interior tissues of the body get their blood under full pressure.

The Lizard Heart

- Lizards have a muscular septum which partially divides the ventricle.
- When the ventricle contracts, the opening in the septum closes and the ventricle is momentarily divided into two separate chambers.
- This prevents mixing of the two bloods.
  - The left half of the ventricle pumps oxygenated blood (received from the left atrium) to the body.
  - The right half pumps deoxygenated blood (received from the right atrium) to the lungs.

Birds and Mammals

The septum is complete in the hearts of birds, crocodiles, and mammals providing two separate circulatory systems:

1. Pulmonary for gas exchange with the environment and
2. Systemic for gas exchange (and all other exchange needs) of the rest of the body.

CIRCULATORY SYSTEM OF FISH

In the diagrams that given below represent the direction of blood flow in blood vessels (arteries and veins). Blood pressure is represented by the thickness of the arrows. Thick arrows indicate high blood pressure. Blood that is rich in oxygen is represented by arrow (Gills to body). Blue arrows represent blood that is low in oxygen after it has passed through the body tissues. Fish have a two-chambered heart with one atrium (A) and one ventricle (V).

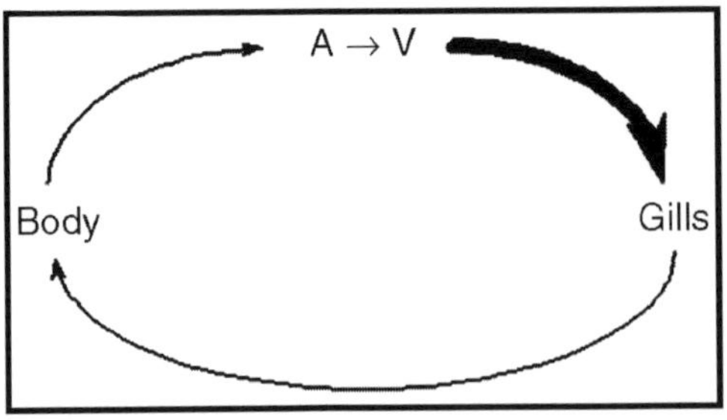

Fig. : ***Blood pressure of Fish***

The gills contain many capillaries for gas exchange, so the blood pressure is low after going through the gills. Low-pressure blood from the gills then goes directly to the body, which also has a large number of capillaries. The activity level of fish is limited due to the low rate of blood flow to the body.

Diseases of the Heart and Cardiovascular System

Cardiac muscle cells are serviced by a system of coronary arteries. During exercise the flow through these arteries is up to five times normal flow.

Blocked flow in coronary arteries can result in death of heart muscle, leading to a heart attack. Blockage of coronary arteries, is usually the result of gradual buildup of lipids and cholesterol in the inner wall of the coronary artery.

Occasional chest pain, angina pectoralis, can result during periods of stress or physical exertion. Angina indicates oxygen demands are greater than capacity to deliver it and that a heart attack may occur in the future. Heart muscle cells that die are not replaced since heart muscle cells do not divide. Heart disease and coronary artery disease are the leading causes of death in the United States.

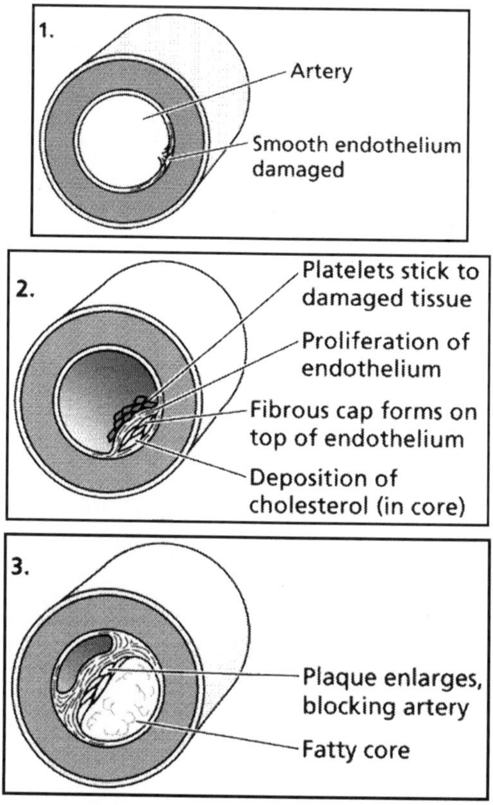

Fig. ***Development of Arterial Plaque.***

Hypertension, high blood pressure (the silent killer), occurs when blood pressure is consistently above 140/90. Causes in most cases are unknown, although stress, obesity, high salt intake, and smoking can add to a genetic predisposition. Luckily, when diagnosed, the condition is usually treatable with medicines and diet/exercise.

The Vascular System

Two main routes for circulation are the pulmonary (to and from the lungs) and the systemic (to and from the body). Pulmonary arteries carry blood from the heart to the lungs. In the lungs gas exchange occurs. Pulmonary veins carry blood from lungs to heart. The aorta is the main artery of systemic circuit. The vena cavae are the main veins of the systemic circuit. Coronary arteries deliver oxygenated blood, food, etc. to the heart. Animals often have a portal system, which begins and ends in capillaries, such as between the digestive tract and the liver.

Fish pump blood from the heart to their gills, where gas exchange occurs, and then on to the rest of the body. Mammals pump blood to the lungs for gas exchange, then back to the heart for pumping out to the systemic circulation. Blood flows in only one direction.

Component of Blood

Plasma is the liquid component of the blood. Mammalian blood consists of a liquid (plasma) and a number of cellular and cell fragment components. Plasma is about 60 per cent of a volume of blood; cells and fragments are 40 per cent.

Plasma has 90 per cent water and 10 per cent dissolved materials including proteins, glucose, ions, hormones, and gases. It acts as a buffer, maintaining pH near 7.4. Plasma contains nutrients, wastes, salts, proteins, etc. Proteins in the blood aid in transport of large molecules such as cholesterol.

Red blood cells, also known as erythrocytes, are flattened, doubly concave cells about 7 μm in diameter that carry oxygen associated in the cell's hemoglobin. Mature erythrocytes lack a nucleus. They are small, 4 to 6 million cells per cubic millimetre of blood, and have 200 million hemoglobin molecules per cell. Humans have a total of 25 trillion red blood cells (about 1/3 of all the cells in the body).

Red blood cells are continuously manufactured in red marrow of long bones, ribs, skull, and vertebrae. Life-span of an erythrocyte is only 120 days, after which they are destroyed in liver and spleen. Iron from hemoglobin is recovered and reused by red marrow. The liver degrades the heme units and secretes them as pigment in the bile, responsible for the colour of feces. Each second two million red blood cells are produced to replace those thus taken out of circulation.

White blood cells, also known as leukocytes, are larger than erythrocytes, have a nucleus, and lack hemoglobin. They function in the cellular immune response. White blood cells (leukocytes) are less than 1 per cent of the blood's volume. They are made from stem cells in bone marrow. There are five types of leukocytes, important components of the immune system.

Neutrophils enter the tissue fluid by squeezing through capillary walls and phagocytozing foreign substances. Macrophages release white blood cell growth factors, causing a population increase for white blood cells. Lymphocytes fight infection. T-cells attack cells containing viruses. B-cells produce antibodies.

Antigen-antibody complexes are phagocytised by a macrophage. White blood cells can squeeze through pores in the capillaries and fight infectious diseases in interstitial areas.

Platelets result from cell fragmentation and are involved with clotting. Platelets are cell fragments that bud off megakaryocytes in bone marrow. They carry chemicals essential to blood clotting. Platelets survive for 10 days before being removed by the liver and spleen. There are 150,000 to 300,000 platelets in each millilitre of blood.

Platelets stick and adhere to tears in blood vessels; they also release clotting factors.

A hemophiliac's blood cannot clot. Providing correct proteins (clotting factors) has been a common method of treating hemophiliacs. It has also led to HIV transmission due to the use of transfusions and use of contaminated blood products.

The Lymphatic System

Water and plasma are forced from the capillaries into intracellular spaces. This interstitial fluid transports materials between cells. Most of this fluid is collected in the capillaries of a secondary circulatory system, the lymphatic system. Fluid in this system is known as lymph.

Lymph flows from small lymph capillaries into lymph vessels that are similar to veins in having valves that prevent backflow. Lymph vessels connect to lymph nodes, lymph organs, or to the cardiovascular system at the thoracic duct and right lymphatic duct.

Lymph nodes are small irregularly shaped masses through which lymph vessels flow. Clusters of nodes occur in the armpits, groin, and neck. Cells of the immune system line channels through the nodes and attack bacteria and viruses travelling in the lymph.

CIRCULATORY SYSTEM OF AMPHIBIANS

- Amphibians have a 3-chambered heart with two atria and one ventricle.
- Blood from the lungs (pulmonary flow) goes to the left atrium.
- Blood from the body (systemic flow) goes to the right atrium.
- Both atria empty into the ventricle where some mixing occurs. The advantage of this system is that there is high pressure in vessels that lead to both the lungs and body.

CIRCULATORY SYSTEM OF SOME REPTILES

In most reptiles, the ventricle is partially divided. This reduces mixing of oxygenated and unoxygenated blood in the ventricle. The partial division of the ventricle is represented by a dashed line below.

CIRCULATORY SYSTEM OF CROCODILIANS, BIRDS, AND MAMMALS

Birds and mammals (also crocodilians) have a four-chambered heart which acts as two separate pumps. After passing through the body, blood is pumped under high pressure to the lungs. Upon returning from the lungs, it is pumped under high pressure to the body. The high rate of oxygen-rich blood flow through the body enables birds and mammals to maintain high activity levels.

TYPES OF CIRCULATORY SYSTEMS

Living things must be capable of transporting nutrients, wastes and gases to and from cells. Single-celled organisms use their cell surface as a point of exchange with the outside environment. Multicellular organisms have developed transport and circulatory systems to deliver oxygen and food to cells and remove carbon dioxide and metabolic wastes. Sponges are the simplest animals, yet even they have a transport system.

Seawater is the medium of transport and is propelled in and out of the sponge by ciliary action. Simple animals, such as the hydra and planaria, lack Specialised organs such as hearts and blood vessels, instead using their skin as an exchange point for materials. This, however, limits the size an animal can attain. To become larger, they need Specialised organs and organ systems.

Multicellular animals do not have most of their cells in contact with the external environment and so have developed circulatory systems to transport nutrients, oxygen, carbon dioxide and metabolic wastes. Components of the circulatory system include

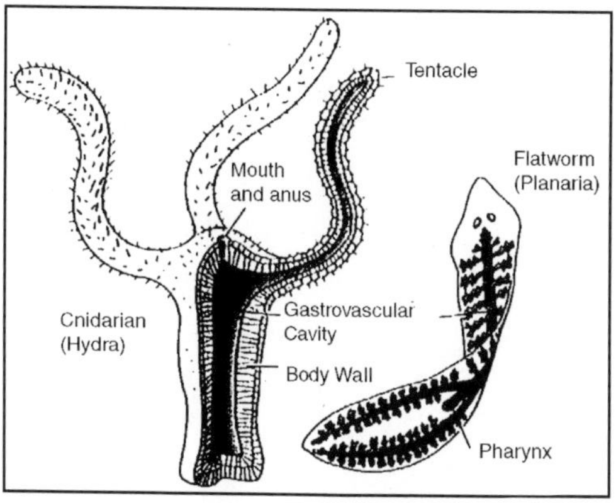

Fig. ***Structures that Serve Some of the Functions of the Circulatory System in Animals that Lack the System.***

- *Blood:* A connective tissue of liquid plasma and cells
- *Heart:* A muscular pump to move the blood
- *Blood vessels:* Arteries, capillaries and veins that deliver blood to all tissues

There are several types of circulatory systems. The open circulatory system is common to molluscs and arthropods. Open circulatory systems pump blood into a hemocoel with the blood diffusing back to the circulatory system between cells. Blood is pumped by a heart into the body cavities, where tissues are surrounded by the blood. The resulting blood flow is sluggish. Vertebrates, and a few invertebrates, have a closed circulatory system. Closed circulatory systems (evolved in echinoderms and vertebrates) have the blood closed at all times within vessels of different size and wall thickness. In this type of system, blood is pumped by a heart through vessels, and does not normally fill body cavities.

Blood flow is not sluggish. Hemoglobin causes vertebrate blood to turn red in the presence of oxygen; but more importantly hemoglobin molecules in blood cells transport oxygen. The human closed circulatory system is sometimes called the cardiovascular system. A secondary circulatory system, the lymphatic circulation, collects fluid and cells and returns them to the cardiovascular system.

Vertebrate Cardiovascular System

The vertebrate cardiovascular system includes a heart, which is a muscular pump that contracts to propel blood out to the body through arteries, and a series of blood vessels. The upper chamber of the heart, the atrium (pl. atria), is where the blood enters the heart. Passing through a valve, blood enters the lower chamber, the ventricle.

Contraction of the ventricle forces blood from the heart through an artery. The heart muscle is composed of cardiac muscle cells. Arteries are blood vessels that carry blood away from heart. Arterial walls are able to expand and contract. Arteries have three layers of thick walls. Smooth muscle fibres contract, another layer of connective tissue is quite elastic, allowing the arteries to carry blood under high pressure.

The aorta is the main artery leaving the heart. The pulmonary artery is the only artery that carries oxygen-poor blood. The pulmonary artery carries deoxygenated blood to the lungs. In the lungs, gas exchange occurs, carbon dioxide diffuses out, oxygen diffuses in. Arterioles are small arteries that connect larger arteries with capillaries.

Vertebrate Vascular Systems

Humans, birds, and mammals have a four-chambered heart that completely separates oxygen-rich and oxygen-depleted blood. Fish have a two-chambered heart in which a single-loop circulatory pattern takes blood from the heart to the gills and then to the body. Amphibians have a three-chambered heart with two atria and one ventricle. A loop from the heart goes to the pulmonary capillary beds, where gas exchange occurs. Blood then is returned to the heart.

Blood exiting the ventricle is diverted, some to the pulmonary circuit, some to systemic circuit. The disadvantage of the three-chambered heart is the mixing of oxygenated and deoxygenated blood. Some reptiles have partial separation of the ventricle. Other reptiles, plus, all birds and mammals, have a four-chambered heart, with complete separation of both systemic and pulmonary circuits.

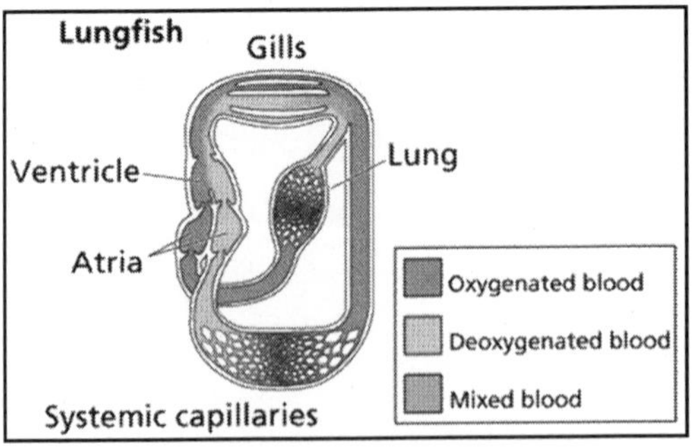

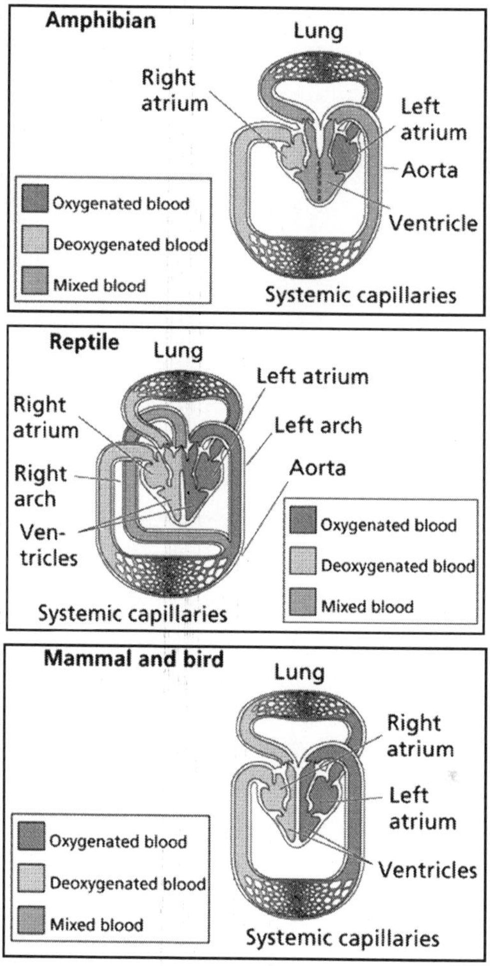

Fig. ***Circulatory Systems of Several Vertebrates Showing the Progressive Evolution of the Four-chambered Heart and Pulmonary and Systemic Circulatory Circuits.***

The Heart

The heart is a muscular structure that contracts in a rhythmic pattern to pump blood. Hearts have a variety of forms: chambered hearts in mollusks and vertebrates, tubular hearts of arthropods, and aortic arches of annelids. Accessory hearts are used by insects to boost or supplement the main heart's actions. Fish, reptiles, and amphibians have lymph hearts that help pump lymph back into veins.

The basic vertebrate heart, such as occurs in fish, has two chambers. An auricle is the chamber of the heart where blood is received from the body. A ventricle pumps the blood it gets through a valve from the auricle out to the gills through an artery. Amphibians have a three-chambered heart: two atria emptying into a single common ventricle. Some species have a partial separation of the ventricle to reduce the mixing of oxygenated (coming back from the lungs) and deoxygenated blood (coming in from the body). Two sided or two chambered hearts permit pumping at higher pressures and the addition of the pulmonary loop permits blood to go to the lungs at lower pressure yet still go to the systemic loop at higher pressures.

Establishment of the four-chambered heart, along with the pulmonary and systemic circuits, completely separates oxygenated from deoxygenated blood. This allows higher the metabolic rates needed by warm-blooded birds and mammals.

The heart beats or contracts approximately 70 times per minute. The human heart will undergo over 3 billion contraction cycles, during a normal lifetime. Thecardiac cycle consists of two parts: systole (contraction of the heart muscle) and diastole (relaxation of the heart muscle). Atria contract while ventricles relax. The pulse is a wave of contraction transmitted along the arteries.

Valves in the heart open and close during the cardiac cycle. Heart muscle contraction is due to the presence of nodal tissue in two regions of the heart. The SA node (sinoatrial node) initiates heartbeat. The AV node (atrioventricular node) causes ventricles to contract. The AV node is sometimes called the pacemaker since it keeps heartbeat regular. Heartbeat is also controlled by nerve messages originating from the autonomic nervous system.

Blood flows through the heart from veins to atria to ventricles out by arteries. Heart valves limit flow to a single direction. One heartbeat, or cardiac cycle, includes atrial contraction and relaxation, ventricular contraction and relaxation, and a short pause. Normal cardiac cycles (at rest) take 0.8 seconds. Blood from the body flows into the vena cava, which empties into the right atrium. At the same time, oxygenated blood from the lungs flows from the pulmonary vein into the left atrium. The muscles of both atria contract, forcing blood downward through each AV valve into each ventricle.

Diastole is the filling of the ventricles with blood. Ventricular systole opens the SL valves, forcing blood out of the ventricles through the pulmonary artery or aorta. The sound of the heart contracting and the valves opening and closing produces a characteristic "lub-dub" sound. Lub is associated with closure of the AV valves, dub is the closing of the SL valves.

Human heartbeats originate from the sinoatrial node (SA node) near the right atrium. Modified muscle cells contract, sending a signal to other muscle cells in the heart to contract. The signal spreads to the atrioventricular node (AV node). Signals carried from the AV node, slightly delayed, through bundle of His fibres and Purkinjie fibres cause the ventricles to contract simultaneously.

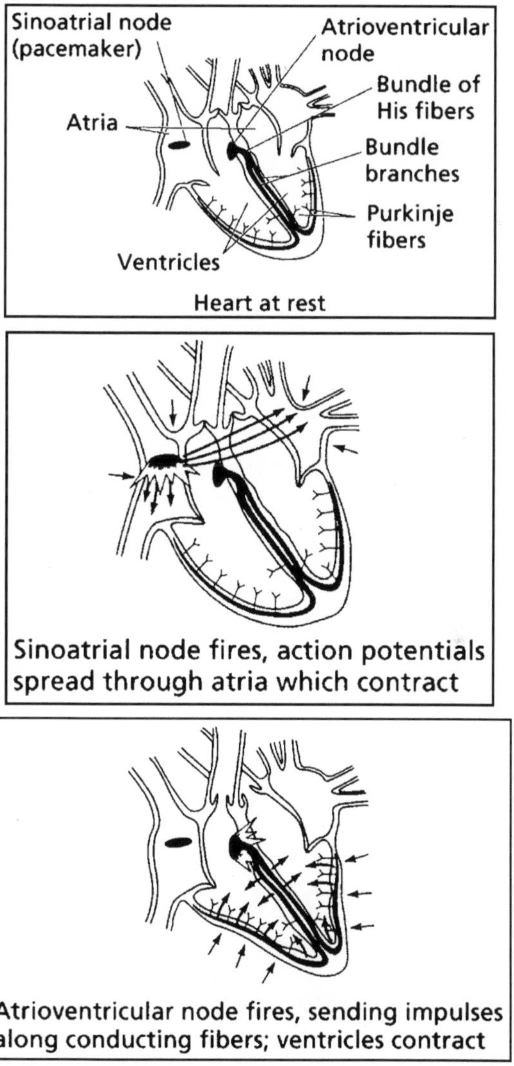

Fig. The Contraction of the Heart and the Action of the Nerve Nodes Located on the Heart.

THE FISH HEART

An efficient circulatory system has:

- A fluid, *e.g.*, blood, to carry the materials to be transported;
- A system of vessels to distribute the blood;

- A pump to push the blood through the system;
- Exchange organs to carry out exchanges between the blood and external environment, *e.g.*,
  - Lungs and intestine to add materials to the blood;
  - Lungs and kidneys to remove materials from the blood.
- The most crucial demand on the circulatory system is the transport of oxygen and carbon dioxide to and from a gas exchange organ:
  - Lungs or
  - Gills and
  - Tissues.
- All exchanges between blood and cells occur in the capillaries.
- The force of the pump that pushes blood through the arteries is dissipated as the blood flows through capillaries. Although capillaries are tiny, the total cross-sectional area of all the capillaries supplied by a single artery is much greater than that of the artery itself. Like a rapid, narrowly-confined stream spreading out over a flat plain, the force and velocity of flow diminish quickly.

This creates a problem:

- If the pump is used to deliver blood with force to the gas exchange organ, little force remains to distribute the oxygenated blood to the tissues.
- If the pump is used to deliver blood with force to the tissues, little force remains to send the deoxygenated blood to the gas exchange organ.

Most fishes have never solved this problem, which is probably why most of them are "cold-blooded":

- □ Blood collected from throughout the fish's body enters a thin-walled receiving chamber, the atrium.
- □ As the heart relaxes, the blood passes through a valve into the thick-walled, muscular ventricle.
- □ Contraction of the ventricle forces the blood into the capillary networks of the gills where gas exchange occurs.
- □ The blood then passes on to the capillary networks that supply the rest of the body where exchanges with the tissues occur.
- □ Then the blood returns to the atrium.

While obviously adequate to the fish's needs, this is not a very efficient system. The pressure generated by contraction of the ventricle is almost entirely dissipated when the blood enters the gills.

8

Animal Diseases

INTRODUCTION

Many of the diseases we are most concerned about are "notifiable" (that is, there are legal requirements on animal keepers). During an exotic disease outbreak our website will provide the latest situation – Defra takes the lead, with partners, in responding to the disease. Vigilance and good stockmanship for all kept animals, including those kept as pets are vital in the fight against animal disease. Monitoring animals for signs of disease, and following good farming and biosecurity practices are essential ways of reducing the risk of disease and preventing the spread of disease during an outbreak.

Notifiable diseases, which if suspected are to be reported to your local office of the Animal Health Veterinary Laboratories Agency (AHVLA), are in an A-Z table linking to information pages about the diseases.

Animal diseases to be notified to your local AHVLA can be "endemic" (those which are already present in the UK, such as Bovine TB), "exotic" (those that are not normally present in the UK, but can be introduced for example via illegal imports or by wild birds), and/or "zoonotic" (those that can pass between animals and humans, such as rabies).

Defra is required under EU law to operate disease control programmes for some diseases (eg *Salmonella* in poultry), and monitor trends for others, in order to minimise the risk to public health.

EFFECT OF ALTERNATE-DAY FASTING ON TYPE 2 DIABETES RISK

Animal Studies

To date, 12 studies using animal models have examined the effect of ADF on chronic disease risk. Approximately half of these studies examined variables related to diabetes, such as fasting glucose and insulin concentrations, fat oxidation, degree of insulitis, and occurrence of type 2 diabetes. Fasting glucose concentrations have generally been reported to decrease in response to ADF in animal models. Three studies found reduced circulating glucose concentrations after a 20–24-wk intervention, whereas one study reported no effect on glucose concentrations after a 16-wk treatment. In the trials that measured insulin concentrations, consistent reductions were noted after ADF regimens that lasted 20 and 24 wk. It is interesting that, in the study of Anson et al, both glucose and insulin concentrations decreased to a similar extent in the ADF and the 40% CR groups. Increases in fat oxidation in liver and muscle have also been observed after relatively short periods (8 wk) of ADF.

Because impaired fat oxidation may contribute to ectopic accumulation of intracellular lipid and the development of insulin resistance, these increases in fat oxidation may increase insulin sensitivity. Also noted by Anson et al was a doubling of the plasma concentrations of ^{2}-hydroxybutyrate in the ADF group but no change in the control group. In contrast, concentrations of this metabolite decreased in the 40% CR group but not in the control group. These results suggest that high rates of fatty acid oxidation leading to ketogenesis occurred with ADF but not with 40% CR. Moreover, reduced occurrence of insulin-dependent diabetes in response to ADF has been reported by Pedersen et al. These authors found that 77% of the BB rats fed ad libitum control diets developed diabetes, whereas only 52% of the animals fasted for 24 h on alternate days became diabetic. The degree of insulitis, however, was not affected, which suggested that the mechanism most likely did not involve modulation of this inflammatory variable.

Human Trials

Risk factors for type 2 diabetes were measured in each of the 3 published human studies of ADF. Evidence from these trials suggests that ADF does not alter fasting concentrations of glucose but may beneficially modulate other indexes of diabetes risk, such as insulin sensitivity. Specifically, Halberg et al observed that, when normal-weight persons fasted for 20-h periods (fast day) and then ate their habitual diet ad libitum on alternate days (feast day), the insulin-mediated glucose uptake increased after 2 wk of intervention, as measured by using the euglycemic-hyperinsulinemic clamp technique. These results are supported by a study conducted by Heilbronn et al, which found that, after 3 wk of ADF, insulin response to a test meal was reduced, which implied improved insulin sensitivity. It is interesting that this effect on insulin sensitivity occurred only in male subjects.

Another diabetes risk factor that has shown a sex-specific effect is glucose tolerance. After 3 wk of ADF, women but not men had an increase in the area under the glucose curve. This unfavorable effect on glucose tolerance in women, accompanied by an apparent lack of an effect on insulin sensitivity, suggests that short-term ADF may be more beneficial in men than in women in reducing type 2 diabetes risk. However, because minimal data and no longer-term studies are available to support this important hypothesis, more studies are needed. The effect of ADF regimens on insulin concentrations appears equivocal. Specifically, Halberg et al found that 2 wk of ADF had no effect on fasting insulin concentrations, whereas Heilbronn et al found that 3 wk of this intervention decreased insulin concentrations, but only after a 32-h fast. Further research examining the time course of ADF effects on such diabetes-related variables could help clarify this matter. Also examined was the responsiveness of skeletal muscle and adipose tissue to ADF. A 2-wk ADF regimen had no effect on intramuscular triacylglycerol (IMTG) concentrations in normal-weight men. In adipose tissue, an inhibitory effect of insulin on adipose tissue lipolysis was observed after 2 wk of intervention. Because increased concentrations of free fatty acids have been implicated in the pathogenesis of type 2 diabetes, this decrease in lipolysis and circulating concentrations of free fatty acids may represent an indirect protective effect of ADF on diabetes risk.

EFFECT OF ALTERNATE-DAY FASTING ON CARDIOVASCULAR DISEASE RISK

As a means of assessing cardiovascular response to ADF, trials in this area have examined heart rate, blood pressure, circulating lipids, and ischemic injury. In a recent study by Mager et al, reductions in heart rate were observed in Sprague-Dawley rats after 16 wk of ADF. Similar effects on heart rate were also found by Wan et al after 24 wk of ADF. In both of these trials, treatment-induced decreases in systolic and diastolic blood pressure were noted after 4 wk, and the lower blood pressures persisted throughout the course of the studies. Moreover, the magnitude of the effect on heart rate and blood pressure was similar in the ADF group and a 40% CR group, which suggests that ADF may be as beneficial as CR in modulating these variables.

In addition, Krizova and Simek observed decreases in circulating lipid concentrations, and, after 8 wk of ADF, both total cholesterol and triacylglycerol concentrations decreased in adult mice. Moreover, the cardiac myocyte response to myocardial infarction induction has been studied by Ahmet et al. MI was induced by coronary artery ligation after 12 wk of ADF or control ad libitum diet in Sprague-Dawley rats. At 24 h after MI induction, the number of apoptotic myocytes in the affected area was one-fourth that in the ad libitum–fed controls, and the size of the MI in the ADF goup was half that in the ad libitum–fed controls. Also noted was a distinct reduction in neutrophil infiltration, which suggested a decrease in inflammatory response.

Heilbron et al examined the effect of ADF on CVD risk. When human subjects fasted on alternate days for a short period (3 wk), circulating concentrations of HDL cholesterol increased, whereas triacylglycerol concentrations decreased. It ispossible that these effects resulted from the decreases in body weight (2.5%) and fat mass (4.0%) observed in these subjects, who were unable to consume sufficient calories on the feast day to maintain an isocaloric state. It is interesting that the shifts in lipid concentrations were shown to be sex specific: ie, only the women had an increase in HDL-cholesterol concentrations, and only the men had a decrease in triacylglycerol concentrations. There is no clear explanation for these sex-based differences. The effect of ADF on blood pressure was also examined in this study. After 3 wk of intervention, neither systolic nor diastolic blood pressured changed in either the male or female subjects. This study included only normotensive persons, however. It will be of interest in future studies to examine the effect of this dietary regimen on blood pressure in other patient groups, such as those who are hypertensive, overweight, obese, and hyperlipidemic.

EFFECT OF ALTERNATE-DAY FASTING ON CANCER RISK: ANIMAL TRIALS

The protective effect of ADF on cancer survival was first described by Seigel et al. In this study, 3–4-mo-old rats were administered an ADF regimen beginning 1 wk before inoculation with MAT 13762 acsites tumor cells. Twelve of the 24 rats (50%) in the diet-restricted group survived 10 d after tumor inoculation, in comparison with only 3 of 24 animals (12.5%) in the control group, which had been fed ad libitum.

The response to ADF of certain biomarkers of cancer risk, ie, insulin-like growth factor-1 (IGF-1), has also been investigated, but the results have been inconsistent. Whereas Anson et al reported increases in IGF-1 in response to 20 wk of ADF, Wan et al reported a clear decrease after 24 wk of treatment. IGF-1 is a potent promoter of cell proliferation and has been shown to decrease in response to CR. The animals in the study of Anson et al did not lose any weight, whereas animals in the study of Wan et al did lose weight; the lack of overall net negative energy balance may explain the different IGF-1 responses in those studies.

More recently, the protective effect of the ADF restriction protocol on age-associated lymphoma and hepatocarcinogenesis was examined in mice. After 16 wk, the incidence of lymphoma in OF1 mice administered an ADF regimen was 0% and that in the control group was 33%. Because the ADF group mice consumed roughly the same total amount of food as the control mice, the efficacy of ADF was independent of total calorie intake. Also noted in that study was a significant, treatment-induced increase in spleen mitochondrial superoxide dismutase (SOD) activity, which was associated with reduced mitochondrial generation of reactive oxygen species (ROS). The effect of ADF on hepatocarcinogenesis has also been examined after 4 wk of ad libitum feeding. In that study, Wistar rats were injected with diethylnitrosamine to initiate liver carcinogenesis and then fed on alternate days for 48 wk. When compared with ad libitum feeding, ADF inhibited the development of preneoplastic lesions and also decreased the number and size of liver nodules. These findings strongly support the hypothesis that long-term ADF may exert an antipromotional effect on experimental carcinogenesis, as has been shown in many studies of CR. Moreover, strong physiologic evidence in favor of the antipromotional effects of ADF was recently reported by Hsieh et al. After 12 wk of treatment, reduced rates of proliferation of several cell types, including mammary epithelial cells, skin epithelial cells (keratinocytes), and splenic T-cells, was observed. These changes induced by ADF were similar to, though not quite as potent as, those seen in the CR groups.

ADF regimens have also been shown to increase mean and maximal life span in certain strains of mice. When ADF regimens were initiated in C57BL/6J mice at ages 2, 6, and 10 mo, body weight decreased and maximal life span was extended. It is interesting that, when the same ADF protocol was administered to A/J mice, body weight was not affected, and life span was increased only in the group that began ADF at age 2 mo. These findings suggest that ADF-induced changes in body weight and maximal life span may be strongly influenced by genotype and the age at which ADF is initiated.

EFFECT OF ALTERNATE-DAY FASTING ON OTHER METABOLIC VARIABLES: HUMAN TRIALS

The effect of short-term ADF regimens on other metabolic variables, such as body temperature, resting metabolic rate (RMR), and various hormone and cytokine concentrations, has also been examined in human subjects.

In the study by Heilbronn et al, neither body temperature nor RMR was affected by 3 wk of treatment. On the other hand, overall fat oxidation was shown to increase by an average of 15 g/d over the course of the trial, according to indirect calorimetry. The authors also observed a positive correlation between fat oxidation and weight loss, which suggested that those subjects with a greater ability to oxidize fat may have lost more weight.

Thus, whether the weight loss noted is a result of ADF may depend on a person's ability to oxidize fat. Heilbronn et al also examined treatment-induced changes in the expression of certain skeletal muscle genes involved in fat oxidation, including 2-hydroxyacyl CoA dehydrogenase, fatty acid translocase, pyruvate dehydrogenase kinase 4, carnitine palmitoy-ltransferase 1, and uncoupling protein 3, as well as the expression of genes implicated in mitochondrial biogenesis, including peroxisome-proliferator-activated receptor-gamma co-activator 1, nuclear respiratory factor 1, and cytochrome C. They reported that a 3-wk treatment had no effect on the expression of any of these genes.

The response of circulating concentrations of certain adipokines—ie, adiponectin, leptin, interleukin 6 (IL-6), and tumor necrosis factor-± (TNF-±)—has also been examined. Concentrations of leptin, IL-6, and TNF-± did not change in response to ADF; in contrast, concentrations of adiponectin increased by 37%. Cnop et al and Higashiura et al both reported that circulating concentrations of adiponectin are positively correlated with insulin sensitivity, which suggests a possible role of this adipokine in the insulin-sensitizing effect noted in the present trial.

SCHMALLENBERG VIRUS

Schmallenberg Virus is a new emerging livestock disease that has been detected in Belgium, Germany, the Netherlands and the UK. It is similar to some other animal disease pathogens, such as Akabane and Shamonda viruses, which are transmitted by vectors, such as midges, mosquitoes and ticks.

The virus has been associated with brief mild/moderate disease (milk drop, pyrexia, diarrhoea) in adult cattle and late abortion or birth defects in newborn cattle, sheep and goats. All the evidence currently suggests that the disease was brought into the UK from infected midges blown across the Channel. We have seen no evidence to suggest that it was from imported livestock. We are closely tracking the disease and will continue to work with partners across Europe and the UK to develop our knowledge of the disease.

Schmallenberg Virus is not a notifiable disease but farmers are asked tocontact their veterinary surgeon if they encounter cases of ruminant neonates or fetuses which are stillborn, show malformations or are showing nervous disease. Veterinary surgeons should then contact their local AHVLA Laboratory

(England and Wales) or SAC Disease Surveillance Centre (Scotland) if they suspect infection with the virus.

BOVINE TB (TUBERCULOSIS)

Bovine tuberculosis (TB) is an infectious disease of cattle and one of the biggest challenges facing the cattle farming industry today, particularly in the west and south west of England. It is caused by the bacterium *Mycobacterium bovis (M. bovis)*, which can also infect and cause TB in badgers, deer, goats, pigs, camelids (llamas and alpacas), dogs and cats, as well as many other mammals.

BIRD FLU (AVIAN INFLUENZA)

Avian influenza (AI), commonly known as bird flu, is a highly contagious viral disease affecting the respiratory, digestive and/or nervous system of many species of birds. It may also pose a threat to people and other animals in certain circumstances.

BLUETONGUE

Bluetongue is a disease of animals affecting all ruminants, including sheep, cattle, deer, goats and camelids (camels, llamas, alpacas, guanaco and vicuña). It does not affect horses or pigs.

AFRICAN HORSE SICKNESS

African Horse Sickness (AHS) is a highly fatal and infectious disease, which affects horses, mules and donkeys. It is carried and spread by biting insects.

PREVENTION AND CONTROL OF ANIMAL DISEASES

People who own or keep animals are responsible for their care and health. The Government has set a number of rules to monitor the health of kept animals. These rules largely concern the prevention, monitoring and control of animal diseases.

The Government has also set specific rules for a number of animal diseases, for example because of their potential social and economic impact or because they pose a threat to human health.

Animal Health Monitoring

The Government constantly monitors animal health in the Netherlands.

This facilitates rapid changes in animal health policy, or immediate response to outbreaks of infectious diseases.

The Government has set three objectives in consultation with the sector:

- To trace outbreaks of known diseases and pathogens that are not endemic (i.e. do not arise in nature) in the Netherlands;
- To identify as yet unknown diseases;
- To monitor trends and developments.

PREVENTION OF ANIMAL DISEASES

To prevent the introduction and dissemination of animal diseases as far as possible, owners of animals must (among other things):

- Do their best to keep animal diseases out of their business;
- Ensure adequate hygiene;
- Be alert to symptoms of disease;
- Comply with requirements when importing animals from other countries;
- Notify a vet of any suspected animal disease.

When there is an increased risk of infectious animal diseases entering the country, more stringent supervision of animal transports is applied.

These measures can be at farm or national level (e.g. a ban on imports from infected countries).

Travellers

There are some holiday destinations where infectious animal diseases, such as foot and mouth or bird flu, are prevalent. The Government therefore advises travellers to be careful about contact with animals abroad. Pathogens from these animals can be accidentally introduced into the Netherlands with travellers and their luggage.

There are also rules for travellers in respect of the importation of animals (in Dutch) andanimal products (in Dutch), such as wool, processed meat or dairy products or hunting trophies. Bringing live animals or animal products into the Netherlands is prohibited.

NOTIFIABLE ANIMAL DISEASES

Stricter rules apply to certain animal diseases in terms of reporting, identification and control. These diseases are on the list of Notifiable animal diseases (in Dutch). When such diseases are notified the Ministry of EL&I can announce additional control measures.

- The reasons for requiring notification of such animal diseases are as follows:
- They can spread quickly;
- They can cause serious damage to the species in question;
- They cannot be prevented or controlled by normal commercial methods;
- They present a serious threat to public health.

Animal diseases can also be designated as notifiable under international conventions.

REPORTING ANIMAL DISEASES

If an animal shows symptoms of a notifiable animal disease, the owner, vet or laboratory must notify the authorities by telephoning the national desk for animal diseases: (045) 546 31 88. This 24-hour service is run by the new Food and Consumer Product Safety Authority (nVWA).

Prompt reporting of an infectious animal disease or its symptoms is crucial. Without early reporting a disease can spread rapidly with potentially massive consequences for humans and animals.

Highly Infectious Animal Diseases

The State is required to combat certain notifiable animal diseases. That means that the Government must always take measures to counter the spread of such diseases. The Minister of EL&I is responsible for this disease control, while the nVWA is responsible for implementation. These animal diseases are treated in accordance with European directives, which are elaborated in the national policy scenarios of the Ministry of EL&I. The approach is regulated in various scenarios.

Supplementary Measures

The Minister for Agriculture and Foreign Trade can take measures over and above the policy scenarios. He is advised by the Chief Veterinary Officer and the Expert Groups.

Experts

There are currently four Expert Groups:

- on ruminants;

- on poultry;
- on pigs;
- on horses.

The experts are independent and are appointed by the Minister. The groups consist of epidemiologists, virologists, vets and (for example) experts in the field of wild flora and fauna.

VACCINATION

There are two types of vaccination:

Preventive Vaccination

Practically all farm animals are vaccinated against diseases. Poultry, cattle, pigs, sheep and goats are all routinely vaccinated against pathogens such as Newcastle disease in poultry and Q fever in goats. This helps to keep the animals healthy and has no harmful effect on their products (meat, milk, etc.).

Vaccination for Disease Control

In the case of certain highly infectious animal diseases such as Classical Swine Fever (CSF) and Foot and Mouth Disease (FMD), emergency vaccination can be used a control measure. This counters the spread of the disease and prevents sickness in animals exposed to the infection. The Ministry of Economic Affairs, Agriculture and Innovation (EL&I) decides in consultation with the EU if and when vaccination should be deployed. The relevant procedure is set out in the Ministry's policy scenarios.

The Rutte Government is campaigning within the EU for emergency vaccination as a control measure during outbreaks of highly infectious animal diseases.

ALTERNATE-DAY FASTING AND CHRONIC DISEASE PREVENTION

Calorie restriction (CR), defined as a reduction in energy intake without malnutrition, has been shown to increase life span, improve numerous functional indexes, and reduce metabolic risk factors for chronic disease in several mammalian species.

CR regimens have consisted of reducing food intake to 60–85% of daily energy needs. As an alternative to traditional CR, another dietary regimen, termed alternate-day fasting (ADF), has also been tested. ADF regimens generally involve a "feast day" on which food is consumed ad libitum that alternates with a "fast day" on which food is withheld or reduced. The feast and fast periods are typically 24 h each, but they may vary. A key point about the ADF approach is that overall calorie intake need not be limited; instead, the frequency of food consumption is altered.

The purpose of this review is to summarize the relatively sparse but highly suggestive literature on ADF regimens. Although the effects of CR on chronic disease prevention were discussed in reviews conducted in the past few years, the ability of ADF to alter chronic disease risk has not yet been summarized. In particular, the key question—whether ADF has effects on risk modulation comparable to those of CR—remains uncertain. Accordingly, our objective was to review the evidence from both animal and human trials concerning ADF and the risk of chronic diseases, such as type 2 diabetes mellitus, cardiovascular disease (CVD), and cancer. In addition, when possible, the magnitude of risk reduction due to ADF will be compared with that due to CR.

BENEFITS OF CALORIE RESTRICTION REGIMENS

A large body of evidence for the physiologic benefits and life-extending properties of CR now exists. Restricting daily energy intake by 15–40% has been shown in both animals and humans to improve glucose tolerance and insulin action, which indicates an enhancement in insulin sensitivity; to reduce blood pressure and the heart rate, which is consistent with benefits for cardiovascular health; and to reduce oxidative damage to lipids, protein, and DNA, which implies a protective effect against oxidative stress.

Many other effects of CR have been documented, including increased average and maximal life span, reduced incidence of spontaneous and induced cancers, resistance of neurons to degeneration, lower rates of kidney disease, and prolongation of reproductive function.

Although the precise mechanisms responsible for such effects are still not clear, several general hypotheses have been proposed—most prominent are the stress resistance hypothesis, the oxidative stress hypothesis, and the induction of a scarcity program hypothesis.

The first hypothesis suggests that, after prolonged dietary restriction, increased resistance to different types of stressors occurs, which permits the cells of many tissues to resist injury induced by genotoxic, metabolic, or oxidative insults. The second hypothesis proposes more specifically that fewer free radicals are produced in the mitochondria of cells, because dietary restriction generally limits energy utilization, which results in less cellular oxidative damage. The third hypothesis proposes that CR induces intrinsic cellular and organismal programs for adaptation to scarcity, which result in the slowing of metabolic processes such as cell proliferation that contribute to senescence; this hypothesis has been strengthened by findings in yeast. The effects of ADF on these proposed mechanisms have not been explored as extensively as have the effects of CR, but some evidence has been generated, and that will be reviewed here.

ANIMAL DISEASE EMERGENCY PROGRAMME

Fighting a disease epidemic or combating other animal health emergencies is in many respects like fighting a war and requires the same level of discipline. It requires the same ability to make rapid decisions based on analysis of the best information that can be made available from all sources, to convert those decisions into clear orders which can be conveyed down the chain to those who are charged with the responsibility of carrying them out and to know that orders have been carried out and with what results. There must therefore be efficient mechanisms in place for the transmission of information and instructions from the national veterinary services headquarters to the front line of the disease eradication campaign in the field and laboratory and for feedback of information to headquarters.

It is clear that for these things to happen quickly and efficiently in an emergency, the veterinary services for a country must be placed in a command structure or line-management system at least for the duration of the emergency response.

National veterinary services are generally structured so as to optimize routine activities such as endemic disease control, veterinary public health, quarantine, etc. In recent years, government veterinary services in many countries have also been rationalized and restructured in many ways, including:

- Regionalization, where the authority and responsibility for delivery of animal health services have been devolved to provinces or regions that match new delegated political structures. This may result in the senior veterinary officer in the region being answerable to an administrative or political superior, who may not fully appreciate the potential national socio-economic consequences of a major animal disease emergency, rather than to the national CVO.
- Rationalization and downsizing of government services, which have led to major retrenchments of professional and technical staff in the public sector, to the point where the remaining staff resources are

inadequate to cope with the major demands of an unexpected animal disease emergency.

- Privatization of veterinary services which has led to the transfer to the private sector of many animal health programmes and functions that have traditionally been the responsibility of governments, including field veterinary services, veterinary diagnostic laboratories and meat inspection.
- Separation of policy functions from operational functions, whereby those arms of government responsible for developing policy and for advising ministers on policy matters are administratively quite separate from those who are operationally responsible for managing major government programmes, including the CVO.
- Separation of veterinary laboratories from the field command. In many countries national veterinary laboratories have been transferred to research administrations, thus weakening links with the CVO and with field veterinary services.

These new structures are frequently not conducive to the mounting of an effective and timely response to an animal health emergency. Countries should review their situation with a view to devising the most appropriate structures and lines of responsibilities that can be rapidly and seamlessly put in place when an emergency arises. This may include organizing one or more of the following well in advance of any emergency:

- An agreement that animal health emergencies will be handled at the national level and that the CVO will assume overall responsibility for responding to the emergency and be directly answerable to the minister in this role.
- An agreement with regional or provincial authorities that their own veterinary staff will come under the line management of the national CVO for an animal health emergency response programme. Arrangements also need to be put in place to ensure that regional field and laboratory veterinary services are fully involved in emergency preparedness planning and training activities and, in collaboration with national veterinary headquarters, in providing early warning of emergencies including emergency disease reporting to national headquarters.
- Similar arrangements for all essential government veterinary services including the central veterinary laboratory to come within the command structure of the CVO, if this not already the case, for the purposes of the emergency response.

- Pre-existing contractual agreements for private sector veterinary organizations, universities and other academic institutions, research institutes, etc. to provide essential services during an animal health emergency.
- Negotiation with the national veterinary association of terms and conditions for hiring practitioners and other private sector veterinarians as temporary government veterinary officers if needed.

CONSULTATIVE COMMITTEE ON EMERGENCY ANIMAL DISEASES (CCEAD)

Countries may find it useful to establish a CCEAD that can be convened as soon as there is a disease or other animal health emergency and that can meet regularly during the course of the emergency response. This would principally be a technical committee whose role would be to review epidemiological and other disease control information, make recommendations on the activation of agreed contingency plans, maintain an oversight over the campaign and advise the CVO and the minister on future plans for the campaign and on implementation of those plans.

A suggested composition of the CCEAD might be:

- CVO (chair)
- director of field veterinary services/director of disease control;
- head of the epidemiological unit;
- directors of state, provincial or regional veterinary services;
- director of the national veterinary laboratory;
- director of any regional veterinary laboratories covering the outbreak areas;
- senior representatives of farmer groups or organizations affected by the outbreak;
- representatives of other key groups, e.g. national veterinary association, universities;
- other technical experts as required (with observer status).

National Animal Disease Control Centre

Countries should establish a permanent national animal disease control centre. In the event of an outbreak of an emergency animal disease, the centre should be responsible to the CVO for coordinating all emergency disease control measures in the country. The centre should be close to the office of the CVO. The epidemiology unit should either be attached to the centre or should work in close collaboration with it. The CVO may delegate day-to-day responsibilities for implementing agreed policy to the head of the centre, who would normally be the director of field veterinary services.

The responsibilities of the national animal disease control centre in the emergency response would include:

- implementing the disease control policies decided by the CVO and CCEAD;
- directing and monitoring the operations of local animal disease control centres;
- maintenance of up-to-date lists of available personnel and other resources, and details of where further resources may be obtained;
- deployment of staff and other resources to the local centres;
- ordering and dispersing vaccines and other essential supplies;
- monitoring the progress of the campaign and providing technical advice to the CVO;
- advising the CVO on the definition and proclamation of the various disease control zones;
- maintenance of up-to-date lists and contact details of risk enterprises, etc.
- liaison with other groups involved in the emergency response, including those that may be activated as part of the national disaster plan;
- preparation of international disease reports and, at the appropriate times, cases for recognition of zonal or national freedom from the disease;
- management of farmer awareness and general publicity programmes, including press releases;
- general and financial administration, including the keeping of records.

The national animal disease control centre should be fully equipped with meeting rooms, a range of maps covering all parts of the country (preferably at 1:50 000), and all suitable communication equipment for liaison with local animal disease control centres, veterinary laboratories, etc., by telephone, radio, e-mail and facsimile, as appropriate. The centre should also be linked with the emergency disease information system.

LOCAL ANIMAL DISEASE CONTROL CENTRES

During an emergency, one or more local animal disease control centres should be set up within easy reach of the infected zones of the disease outbreak. Ideally they should be sited so that teams are able to travel to and from any site for surveillance or any other disease control activities on the same day. Where distances are not great, these local centres could be established on a permanent basis in a regional or district veterinary or agricultural office. Otherwise, possible locations for temporary local disease control centres (e.g.local government offices) should be identified and agreed in advance.

The local animal disease control centre should be fully equipped with offices, meeting rooms, maps, communication equipment to contact both field personnel and the National Animal Disease Control Centre, vehicles and fully stocked central stores. Central cold-storage facilities for vaccines should also be located at or within easy access of the centre. The centre should have simple equipment that will allow it to process and dispatch diagnostic specimens, including serum samples.

Each local animal disease control centre should be under the control of an experienced senior field veterinary officer. This officer should be given the responsibility for directing the emergency disease control and eradication programme within the area, under the general supervision of the national animal disease control centre and the CVO. All staff allocated to a centre for the period of the disease emergency should be under the command of this field veterinary officer for the duration of their attachment. The officer in charge of the centre should be given the authority to:

- designate a farm, herd or community as an infected premises, when necessary, after consulatation with, and with the agreement of, the national animal disease centre;
- quarantine infected and dangerous contact premises;
- send surveillance teams to all places where there are susceptible livestock;
- deploy the necessary staff to infected premises to arrange valuation, slaughter and safe disposal of animals, cleaning and disinfection;
- advise on the delineation of infected, surveillance and control zones, and on the measures to be taken in them;
- impose livestock movement restrictions;
- suspend the operations of, or place zoosanitary restrictions on, livestock markets, abattoirs and other risk enterprises;
- organize and implement vaccination programmes;
- carry out insect vector control programmes, if necessary;
- liaise with police and other authorities over the maintenance of disease control restrictions;

- liaise with local wildlife authorities;
- carry out publicity campaigns;

Zoning is An Integral Part of Disease Control and Eradication Campaigns

The local animal disease control centre should be allocated sufficient staff to carry out these functions properly. Each major area of field activity should be under the control of an experienced veterinary officer. The centre should also have a veterinary epidemiologist, who can provide specialized advice to the officer in charge and take care of disease reporting and the emergency disease information system.

Depending on the type of disease control strategy chosen, there will be a need for disease surveillance teams, vaccination teams, quarantine and livestock movement control staff, valuers, infected premises teams (livestock slaughter, disposal, cleaning and disinfection), administrative staff (stores and general administration) and a public relations/education officer.

DIFFICULT OR MARGINALIZED AREAS

Countries may be faced with the situation where they have to deal with an outbreak of an epidemic livestock disease in areas that are difficult for geographical reasons or because they are relatively inaccessible owing to civil unrest or because they practise nomadism or transhumance.

Such areas frequently have little contact with outside government officials. The conventional approaches recommended above will need to be considerably modified in these circumstances.

Only staff experienced in the local conditions and who can gain the confidence of local communities should be used.

Sometimes the main outside contacts of such communities will be through agricultural and other specialists employed by non-governmental organizations (NGOs). NGOs and their staff should be regarded as a valuable resource for assistance in implementing animal health programmes in difficult areas, including epidemic livestock disease control campaigns.

Negotiations should therefore be carried out with appropriate NGOs to obtain their collaboration in this area. The necessary training and resources should then be supplied to their staff.

Community animal health workers are another valuable resource. Their help should be enlisted and they should be suitably trained and equipped.

ANIMAL DISEASE EMERGENCY PREPAREDNESS PLANNING

Responsibility for Animal Disease Emergencies

The Chief Veterinary Officer (CVO) or equivalent, such as the Director of Veterinary Services of the country, should have overall technical responsibility with regard to preparedness for and management of animal health emergencies. The appropriate government minister would of course be ultimately responsible.

Responsibility for Animal Disease Emergencies with a Public Health Component

Animal disease emergencies that have a significant public health component are a special case. These emergencies might occur, for example, in a major outbreak of a zoonotic disease such as Rift Valley fever, Japanese encephalitis, Venezuelan equine encephalitis or rabies. For these emergencies, negotiations should be carried out between the Ministry of Agriculture and the Ministry of Health (or their equivalents). Agreement should be reached in advance on a joint framework for preparing contingency plans and for other complementary preparedness programmes. Agreement should also be reached on the most efficient mechanisms for coordinating emergency responses, for implementing disease control and eradication programmes and for sharing responsibilities. Appropriate opportunities for sharing resources between the two agencies should also be explored so as to avoid unnecessary duplication. This might include a single diagnostic laboratory facility for the zoonosis in question, or at least the sharing of diagnostic reagents and of expertise between government veterinary and medical laboratories, common cold-chain facilities for vaccines, joint field missions and joint public awareness and public relations campaigns.

Of critical importance is the development of coordinated and efficient mechanisms for the rapid exchange of emergency disease reports and other key epidemiological information between the two agencies. These arrangements should apply at local and regional levels as well as at the national headquarters of both ministries. This is vital in order to enable a rapid response to new disease incidents and extensions of the outbreak, whether they are first manifested in humans or animals.

GETTING STARTED-OBTAINING SUPPORT

In order to have emergency preparedness planning recognized as an important core function of national veterinary services, and to have adequate funding and other resources allocated to these activities, the CVO should enlist the support of all interested parties.

These would include, *inter alia,* the CVO's own minister and senior ministry officials, other government departments and agencies including national economic development planning authorities, farming communities and organizations, livestock marketing authorities, livestock traders and exporters and livestock product processors.

Of these, the most important target groups are the government and the farming community. In presenting a strong case for support for emergency preparedness planning, the identified risks of the transboundary animal disease or other animal health emergency, and analysis of those risks, should be described together with the potential socio-economic consequences of an incursion or epidemic of the disease.

Additionally, the benefits that will result from more rapid containment and eradication of the disease outbreak through forward contingency planning and preparedness should be forcefully presented. The case should preferably be supplemented by a formal socio-economic cost-benefit analysis.

NATIONAL ANIMAL DISEASE EMERGENCY PLANNING COMMITTEE

A National Animal Disease Emergency Planning Committee (NADEPC) should be appointed to facilitate and coordinate emergency planning. This committee should be directly accountable to the Minister of Agriculture and should be charged with the responsibility for developing and maintaining a high state of preparedness for animal disease emergencies. It should preferably be chaired by the CVO and should hold regular meetings to carry out the following functions:

- commissioning of risk assessments on high-priority disease threats and subsequent identification of those diseases whose occurrence would constitute a national emergency;
- appointment of drafting teams for the preparation, monitoring and approval of contingency plans and other documents;
- liaison with, and involvement of, relevant persons and organizations outside the government animal health services who also have a role in animal health emergency preparedness planning. This would include, *inter alia,* the national veterinary association, livestock industry groups, the national disaster management authority and departments of finance, health and wildlife;

- enhancement of the capabilities of emergency field and laboratory veterinary services, especially for specific high-priority livestock disease emergencies;
- development of active disease surveillance and epidemiological analysis capabilities and of emergency reporting systems;
- staff training and farmer awareness programmes;
- ssessment of resources needs and planning for their provision during animal health emergencies;
- drafting of legislation and development of financial plans;
- implementation of simulation exercises to test and modify animal health emergency plans and preparedness;
- overall monitoring of the national state of preparedness for animal health emergencies.

NADEPC should comprise the CVO as chair, the national animal disease planning officer as secretary, director of field veterinary services/director of disease control (or equivalent), director of the national veterinary laboratory, head of the epidemiological unit, director of animal quarantine and directors of state or provincial veterinary services. In addition to these senior animal health officials, representatives of other ministries that may have a substantial role in responding to animal health emergencies, such as health, wildlife services, economic planning and finance, should either be full members of the committee or should be coopted as required. It is also highly desirable to have members drawn from the private sector, such as representatives of major livestock farming and processing organizations.

National Animal Disease Emergency Planning Officer or Unit

A National Animal Disease Emergency Planning Officer should be appointed. This officer should be a senior veterinary officer with training in epidemiology and wide field experience in the management of disease control programmes. If circumstances warrant it, a small unit of professionals should also be appointed. The planning officer would be both the adviser to and the executive officer of the National Animal Disease Emergency Planning Committee, and would be actively involved in all NADEPC programmes itemized above.

ANIMAL DISEASE EMERGENCIES AS A COMPONENT OF THE NATIONAL DISASTER PLAN

Most countries have well-developed national disaster plans. These allow essential government and non-government services and resources to be rapidly mobilized in response to a disaster. Such plans may also allow these essential services to be given special powers to act in the emergency. The national disaster plan is usually aimed at specific natural disasters of an emergency nature such as major fires, floods, hurricanes, earthquakes and volcanic eruptions.

A strong case can be made for the official recognition of a disease emergency as a defined natural disaster situation which can be incorporated into the national disaster plan. An epidemic of a transboundary animal disease, for example, has the same characteristics as other natural disasters: it is often a sudden and unexpected event, has the potential to cause major socio-economic consequences of national dimensions and even threaten food security, may endanger human life and requires a rapid national response.

There are several essential government services, other than the Ministry of Agriculture, which will be invaluable in an emergency. These include, *inter alia*:

- defence forces (notably the army and air force) which can provide support for such activities as transportation of personnel and equipment to disease outbreak sites, particularly when these are inaccessible to normal vehicles; provision of food and shelter; protection of disease control staff in areas with security problems; and provision of communication facilities between national and local disease control headquarters and field operations;
- police, for assistance in the application of necessary disease control measures such as enforcement of quarantine and livestock movement and protection of staff if necessary;
- public works department, for provision of earth-moving and disinfectant-spraying equipment, and expertise in the disposal of slaughtered livestock in eradication campaigns;
- national or state emergency services for logistical support and communications.

Once approval has been given for the recognition of animal health emergencies within the national disaster plan, a set of standard operating procedures should be prepared and agreed with all cooperating agencies. The format of these documents will presumably be determined by pre-existing arrangements for the national disaster plan. They should set out in simple, unambiguous terms just how the national disaster plan is going to be activated in the case of an animal health emergency.

They should also describe what duties and functions the support agencies may be expected to perform under different circumstances. Finally, they should establish the formal relationship between the various agencies and the chain of command. It should be emphasized that the Ministry of Agriculture (or equivalent ministry responsible for animal disease issues) is the lead combat authority during the emergency response.

ALTERNATE-DAY FASTING IN HUMAN AND ANIMAL TRIALS

Some discrepancies between human and animal ADF data are evident. With regard to the effect of ADF on the risk of type 2 diabetes mellitus, the results to date from human trials have been inconsistent, whereas the animal evidence suggests favorable alterations. Fasting glucose concentrations in rodents, for example, consistently were lower after 20 wk of treatment, whereas, in humans, no effect was seen after 2–3 wk of ADF. It may be that longer intervention periods are required to alter glucose concentrations in human subjects. In the case of fasting insulin concentrations, equivocal findings were noted in humans, whereas consistent decreases have been seen in animals. Moreover, animal data indicate that ADF is just as efficacious in decreasing fasting glucose and insulin concentrations as is daily CR. Although neither glucose nor insulin concentrations were affected in the brief human trials carried out to date, it is interesting that findings such as increased insulin-mediated glucose uptake and reduced adipose tissue lipolysis have been reported. In animal models, fatty acid oxidation in liver and muscle is increased with ADF; this may indicate that ectopic accumulation of intracellular lipid could be decreased, which in turn may lead to improved insulin sensitivity. Nevertheless, it should be noted that, in obese human subjects, CR does not affect intramyocellular lipid content but does result in decreased lipid accumulation in liver. Complementary to this evidence, the incidence of type 2 diabetes risk was lower in rodents fed on alternate days than in ad libitum–fed controls. In sum, the favorable effects noted in animal studies suggest that prolonged ADF is a beneficial means of lowering type 2 diabetes risk. Results from human studies, however, are less clear. It seems reasonable to expect that ADF will improve insulin sensitivity in humans, but the conflicting findings make it difficult to be certain about this. Longer intervention trials in human subjects may help to clarify this issue.

Although overall beneficial modulations in risk factors for vascular disease have been found, with respect to blood pressure, the evidence from animal studies has shown a consistent decrease in both systolic and diastolic readings, whereas data from human trials have shown no effect on either variable. One possible explanation for this inconsistency may be the differences in the duration of intervention. Because an effect on blood pressure readings was identified only after a 4-wk treatment in animals, treatment for 3 wk in the human study may not have been long enough. Alternatively, weight loss, ie, negative energy balance, may be required for a blood pressure effect in humans. In the case of circulating lipid concentrations, beneficial modulations have been noted in both human and animal studies. Specifically, human data show treatment-induced increases in HDL-cholesterol concentrations and reductions in triacylglycerol concentrations, and results in rodents show decreases in total cholesterol and triacylglycerol concentrations. Also shown, although only in rodent models, were improvements in cardiac response to MI induction and decreases in heart rate. In addition, the decreases in heart rate and blood pressure induced by ADF were similar to those induced by CR. Taken together, these improvements suggest that ADF may help reduce the risk of CVD.

To date, the direct effect of ADF on cancer has been tested only in animal models. Most of those trials suggest a pronounced beneficial effect on cancer risk factors, including substantial decreases in lymphoma incidence, increases in spleen SOD activity accompanied by reductions in ROS generation, inhibition of hepatic preneoplastic lesion development, and a greater survival rate after tumor inoculation. The physiologic evidence of clear reductions in proliferation rates of several cell types—including mammary epithelial cells, keratinocytes, and splenic T-cells—induced by ADF regimens also supports the antipromotional actions of this intervention. The antiproliferative mechanism remains unknown, however, because the effects of ADF on IGF-1 concentrations have been inconsistent. Nevertheless, most of these studies have reported a protective effect, so it is reasonable to propose that ADF will prove to be an effective means of decreasing cancer risk. Studies in human subjects are still required to answer this important question.

An interesting but unresolved issue is the effect of ADF on body weight. Body weight has been shown to be highly variable in response to ADF in both human and animal models. In some animal models, when ADF regimens have been applied in the short term, no effect on body weight has been noted after 2 wk, whereas gains in weight were noted in other trials after 8 wk. It is possible that the animals in the 8-wk trial may have overcompensated for the lack of food on the fast day by eating more than twice their average daily intake on the feast day. In other studies, when ADF regimens were administered for 12 wk, body weight was found to decrease, but, when ADF regimens were administered for 16 wk, no effect on body weight was observed. Trials examining the effects of long-term ADF in animals have fairly consistently found decreases in body, although the study by Anson et al did not. Such findings suggest that the animals were unable to consume twice their daily food intake on the feast day for longer periods, which resulted in a loss of body weight. There clearly is variability in the capacity of animals, even within the same strain of mouse, to compensate for a fast day on the feast day. A variety of factors, such as housing conditions, palatability, or energy density of diet, and genetics can be hypothesized as influencing compensation. Understanding the factors controlling compensation is an important area for future research.

The effect of ADF on body weight in humans is difficult to infer because of the very short trial durations and the small number of studies published to date. As was seen in animal trials, 2 wk of ADF had no effect on body weight in normal-weight human subjects. Nevertheless, when the intervention period in humans was extended to 3 wk, a decrease in body weight was noted. This decrease in body weight may have resulted from an inability to consume an adequate amount of food on the feast day to sustain body weight. Similar findings were noted in animals, but only after much longer trial durations (>20 wk). An important study design issue is whether weight loss should be prevented in human ADF studies by "forced" maintenance of calorie intake. Forcing maintenance of intake may produce harmful effects over the long term, so this approach cannot currently be recommended. It is also possible that weight loss in humans following an ADF regimen will prove to be minor or transient. In contrast, striving for full compensation of calorie intake (ie, no weight loss) in studies lasting only weeks or a few months is unlikely to have significant adverse health consequences. The absence of weight loss would allow useful comparisons and distinctions between ADF and CR. These issues will require consideration by investigators who conduct future human trials with ADF.

Findings to date from both human and animal experiments indicate that ADF may effectively decrease the risk of CVD, whereas results from animal studies suggest a protective effect on cancer risk. In terms of diabetes prevention, animal data suggest a beneficial effect, but human data have been equivocal. However, it is important to note that the human studies examined in this review are limited; they all lacked control groups and used short trial lengths. Future studies with longer trials and including control groups are needed to answer these important questions. The effect of ADF regimens in insulin-resistant or diabetic populations also should be determined, because they could help to clarify the role of ADF as a treatment for preexisting diabetes rather than as a protection against diabetes. Moreover, human ADF trials in modestly overweight persons, who are at greater risk of chronic disease, are warranted. In this context, it is important to note that the control animals in both the CR and ADF studies are likely to have been obese, because they were fed ad libitum.

ADF regimens also may be as efficacious as daily CR in improving certain indexes of risk of type 2 diabetes and CVD, although the number of studies directly comparing the 2 regimens is small. Further analysis of the mechanisms responsible for beneficial effects of ADF is clearly warranted, particularly if these effects occur in the absence of negative energy balance. Novel mediators and therapeutic strategies may thereby be uncovered. Finally, it seems intuitively likely that persons will find it easier to fast or reduce intake on alternate days than to reduce their intake every day. For this reason, ADF regimens may allow better compliance than would CR regimens and may represent an attractive area for investigation.

It will also be important to understand whether the mechanisms by which ADF protects against chronic disease risk are similar to those of CR. Indirect evidence suggests that the 2 regimens may share mechanisms. For instance, the study of Descamps et al reported increases in spleen mitochondrial SOD activity accompanied by decreases in mitochondrial generation of ROS as a result of ADF. Such findings suggest that ADF may act by increasing resistance to oxidative insult, which is a key feature of the stress resistance hypothesis.

In summary, this still nascent literature suggests that ADF may effectively modulate metabolic and functional risk factors, thereby preventing or delaying the future occurrence of common chronic diseases, at least in animal models. The effect of ADF on chronic disease risk in normal-weight human subjects remains unclear, however, as do the mechanisms of action. Much work remains to be done to understand this dietary strategy fully.

9

Hormones in Animal Production

Hormone-dependent sex differences in growth rate have been known for a long time. It has also been known that growth rate and FCE (feed conversion efficiency) are higher in intact males than in castrates. It was natural, then, that the availability of hormones and other natural or synthetic substances displaying hormonal activity led to experiments aiming at their use to increase production. Beginning in the mid-1950s, DES (diethylstilboestrol) and hexoestrol were administered to cattle increasingly in the US and the UK respectively, either as feed additives or as implants, and other types of substances also gradually became available.

In general, such treatment has resulted in 10–15% increases in daily gains, similar improvements in FCE and improvement of carcass quality (increased lean/fat ratio). Thus there has been a substantial reduction in the amount of energy required per unit weight of protein produced, and the economic implications of this have been great. While the use of hormonally active substances in animal production rose, opposition to their use also increased, because of the theoretical possibility that residues in edible tissues might endanger consumers. The factors leading to the ban on DES in the US, first imposed in 1973, have been described.

Several reports confirm that DES endangers the health of animals and man, when repeatedly used in large doses . However, as regards risks due to the presence of residues in meat produced according to regulations, no documented deleterious effects have ever been reported in man, either from DES or any other substance with hormonal activity. A distinction should be made between the hormones as such, for which the metabolism in the body is relatively well known, and synthetic or other substances for whose metabolic inactivation the body may not possess the enzymes necessary. When natural hormones are used in animal production, claims of zero-tolerance residue levels are not meaningful, since these compounds occur in detectable and highly variable concentrations in body fluids as well as in the tissues of all animals, treated or not . For other substances with hormonal activity the situation is different. However, when residue levels are extremely low, it seems reasonable to weigh the potential risks against the undisputed positive effects some of these compounds have in animal protein production.

This paper will discuss types of substances with hormonal activity currently in use or under investigation, their effects, mechanism of action, metabolism/elimination, tissue levels, risks to the consumer and their economic importance. Finally, other avenues to increased animal production as alternatives to use of hormones will be briefly envisaged. For the sake of simplicity the term hormone will be used, even if incorrectly, to cover all substances with hormonal activity, whether natural or synthetic. Since much information on the question collected before 1975 has been reviewed previously, the main emphasis will be placed here on research since that time.

USE OF HORMONES IN ANIMAL PRODUCTION

A reading of many popular, and even official, papers reveals that different types of anabolic agents, whatever their structure, are grouped together under the general name of hormones, among them those whose activity as oestrogens is emphasized and those which reinforce masculine traits. Published material and research results make it possible to distinguish three types of anabolic agents. The first type includes those not found in mammals and birds but obtained exclusively through organic synthesis: diethylstilboestrol (DES) or stilboestrol, dienoestrol, hexoestrol and their derivatives. These compounds, active when administered orally or as implants, are harmful for laboratory animals, whether given directly or as residues in the flesh and offal of treated animals. The use of these substances, a potential danger to human health, is prohibited in many countries, particularly since the methods of measuring DES levels in meat are unreliable, so that measurements must be made on the excreta. *The second type of anabolic agents consists of the natural steroids, normally present in all animals*: testosterone, oestradiol, oestrone, progesterone. Secreted by the endocrine glands, these substances and their derivatives, although absorbed by the intestines, are deactivated in the liver. They produce measurable effects only when implanted.

The third type, found in the higher plants and moulds, consists of substances with a more or less clear oestrogenic and, at times, anabolic activity. Among these phytoestrogens, only zearalenone has been isolated for use in animal production: as zeranol, it gives good results, but has an effect on genital formation and the level of thyroxine in plasma. It should be added that Sharaf and Gomaa attribute oestrogenic effects to Vitamins E, B_6, C and perhaps A, while Nelson *et al*. have studied the same effects in O-p′-DDT. This paper discusses only the consequences for the consumer of the use in animal production of the first two types: the synthetic agents such as DES and the natural steroids oestradiol, progesterone and testosterone. First mention should be made of cases of cancer of the vagina and hypogonadism observed in the offspring of women treated with DES.

Such observations are verified by research based on the so-called relay methodology, based on the fact that methods of analysis for a residue and its metabolites fail to reflect practical realities, since they may lead to findings of residues without bioavailability or *vice versa*. Further, the individual biological evaluation of each metabolite fails to reflect the overall consumption of all of them. The relay, a farm animal, changes this situation. In order to determine whether this was true of the first two types of anabolic agents, rats and mice of both sexes received balanced diets containing 20% of the flesh or 6% of the liver of male and female calves implanted with DES (24 mg in each of two implants), oestradiol and progesterone (20 + 200 mg in two doses) or oestradiol + testosterone (20 + 200 mg in two doses). Animals were implanted at the beginning of the growth period and 60 days later, 38 days prior to slaughter. Chemical measurement of residues showed DES levels of 30 g/kg in the flesh, but none in the liver. Levels of natural steroid hormones did not exceed levels in untreated animals. Feeding rats and mice on the flesh of DES-implanted calves led to growth in both sexes, sterilized the females and sharply atrophied the testicles, seminal ducts and penis of the males. Liver behaviour was less clear. Pokrovski *et al*. and Nesterin obtained similar results with flesh. It was also observed that DES persisted for nearly 40 days in the environment and was found in alfalfa.

Animals fed on the flesh and liver of calves implanted with oestradiol + progesterone or oestradiol + testosterone behaved in the same manner as controls fed on the flesh and liver of control calves. After 24 months, the percentage of tumors (primarily mammary adenomata, frequent in old Wistar rats) was the same in both groups. Trenbolone acetate (TBA), a testosterone derivative, leaves little or no residue. Measured by radio-immunoassay, Gropp *et al*. also studied their biological effects by the relay toxicity method, on Sprague-Dawley rats fed on the flesh of calves implanted with a mixture of 20 mg oestradiol-17β + 140 mg TBA, as well as with doses 10 and 25 times higher. The results of this study, which continued for 110 weeks and covered two generations, showed the absence of any unfavourable influence on growth, organ weight, reproduction, teratogenicity and cancerogenicity. Only females fed on the flesh of control calves to which TBA had been added in doses 25 times normal, had significantly lower growth($P < 0.05$). The biochemical constants in the blood were the same in both groups.

These experiments show that compounds of natural steroid male and female hormones, and their derivatives, appear to be without danger for the consumer. They are useful in animal production, obviating a number of often costly and polluting therapeutic operations. On the contrary, the use of DES-type non-steroid anabolic agents has serious disadvantages. It is difficult to detect their residues in meat. Relay toxicity results show that analyses and biological behaviour fail to agree. The indiscriminate prohibition of all types of anabolic agents may be leading to the use of DES and similar substances. Analyses of urine in which they can be identified most easily, have revealed them in 20% of cases. In the United States, McClung estimates that 500 000 bovines are being treated illegally. In Europe, illegal use appears to be fairly general, but is tending to decrease. The use of DES and similar substances should thus be prohibited, while that of the steroids and their derivatives should be authorized under veterinary control. This was the view of many writers as early as 1967, and their conclusions were confirmed by the FAO/ WHO Joint Committee of 1975 and a recent international symposium. Any new anabolic agent should be severely tested before it is distributed widely. There is a need to develop practical, economic and rapid — but very exact — methods of detecting residues, particularly those originating in products chemically different from the natural steroid hormones and their derivatives.

METHODS OF MEASURING HORMONE

It should be clearly understood that this discussion concerns only measurement methods intended to detect residues remaining in flesh and viscera. These methods are, however, not very reliable, and other biological material must be examined to determine whether the animals have been treated. Specialists agree that there are still difficulties in testing for residues in the tissues of animals intended for use as food. Many different methods have been proposed and discussed at the EEC level. While the trend is increasingly towards the use of radio-immunoassay, it is undeniable that the techniques are highly complex and still far from being perfected. DES and similar substances should preferably be looked for in urine and faeces. Instructions to this effect were issued in the United States in 1979 and confirmed in 1980. Account should also be taken of the fact that no difference may be found between animals properly implanted with anabolic agents and untreated animals.

This problem has received insufficient attention. The proceedings of the FAO/WHO Joint Meeting of March contain reports emphasizing this lack of difference and the relatively high, but physiological, levels of testosterone and progesterone in the flesh of bulls and gestating cows, respectively. This being the case, how can the natural presence of these substances be distinguished from their presence due to the use of natural products? Further, it should not be forgotten that very many animal feeds contain phytohormones which can cause modifications in the morphology and histology of certain organs. After summarizing and stating the principles of the four main groups of useful techniques enumerated by Kroes *et al.* and Richou-Bac and Pantaléon—histological, biological, physico-chemical and radio immunoassay methods—this paper will study in detail the techniques that can be applied in practice for routine control.

- Histological methods, used specially in young bovines, consist in the observation of modifications in the male and female genitalia. The prostate and the bulbo-urethral glands are particularly examined in male calves and lambs, while in females attention is focused on the glands of Bartholin, the teats, the vagina and the cervix. Modifications are obvious when DES and zeranol, as well as hexoestrol, have been used. In pigs the effects are less discernable. It is generally agreed that the histological method makes it possible to conclude that an oestrogen has been used at some earlier time. The use of natural steroid oestrogens, in combination with testosterone or progesterone, leads only to slight histological modifications. The phyto-oestrogens may also have some effect and lead to modifications; this is the case of Fusariuminfected maize.
- The biological methods used are those of Astwood, which measures the increase in uterus weight of pre-puberal mice and rats, and of Allen and Doisy, based on modifications of the cytology of the vagina of adult mice after ovarectomy. The latter test, applied by Stob *et al.* in 1954, is the most sensitive, but it is responsive only to compounds with oestrogenic activity. Martin reports that the use of concentrated extracts placed in the vagina of mice also gives good results.
- Physico-chemical methods, which make it possible to identify with precision the anabolic agent sought for, require a fairly long extraction, followed by thin-layer or gaseous-phase chromatography. Hans and Abraham report that the latter techniques, combined with mass spectometry, can now be contemplated. Sensitivity is high. Further off is the possible use of liquid/liquid chromatography. Thin-layer

chromato-graphy is currently used, and Verbeke gives details in an EEC document, while Waldschmidt and Schuller and Stephany have described the method.

- Radio-immunoassay, first described for the oestrogenic steroids by Jiang and Ryan, is based on competition between tritium-labelled and unlabelled antigens for the corresponding specific antibodies. This technique, increasingly used, is highly sensitive, making it possible to detect 0.05 to 0.005 ppb of DES and 0.1 ppb of TBA. It is, however, complex and long, although it is being improved. It has the same advantages as all uses of radioactive compounds. The technique has been well studied by Hoffmann and Hoffmann and Karg.

RESIDUES IN EDIBLE TISSUES OF HORMONE-TREATED ANIMALS

Much work has been devoted to the development of sensitive methods of detecting hormone residues in meat from hormone-treated animals. As regards compounds given orally, it should in principle be possible to realise claims of zero-tolerance residue levels, by selecting the proper withdrawal time. During recent years, the use of implants has, however, gained in importance. While removable implants have been tested in steers, with no decrease in performance when withdrawn 32 and 39 days before slaughter, the wide use of non-removable implants makes residue studies important. Determination of normal levels of endogenously produced natural hormones is also important, to enable risk evaluation to be carried out in realistic terms.

Several residue studies have been made of synthetic as well as natural compounds, mainly in cattle. When regulations governing dose, sites of implantation and timing in relation to slaughter are adhered to, residue levels of DES, hexoestrol and oestradiol-17β in edible tissues have generally been in the lower ppb to the ppt range, *i.e.* from a few ng/g down to some hundred pg/g of tissue. In the latter case there was almost complete overlap between values for untreated and treated steers after 105 days. Zeranol implants have so far not left detectable residues in edible tissue.

Most studies of androgens have concentrated on TBA. The ester being rapidly hydrolysed, measurements of residues have been limited to the free compound and/or its major metabolite. Results based on radio-immunoassay of extracts or on radioactivity measurements have indicated levels in edible tissue of the order of 1 ppb or below. In a recent study using implants containing tritiated TBA in heifers, it was found that when slaughter took place 60 days after implantation, the major proportion of tritium-containing residues was not extractable with organic solvents. In muscle 95.5%, in liver 94.4%, in kidney 98.8% and in fat 59.1% of the radioactivity remained in the aqueous phase, not quantifiable by radio-immunoassay. This suggests that the major part of the residues after TBA implantation occurs in a non-extractable, possible covalently bound form in tissues.

Residue levels of gestagens have been also measured, in connection with their use as growth stimulants. Residues of melengestrol acetate used as a feed additive in daily doses of 0.25 to 0.50 mg per head have consistently been below the sensitivity levels of the methods used (*i.e.*, below 10 ppb in fat, liver, muscle and kidney), whether or not the compound was withdrawn 48 hrs before slaughter. According to the Agricultural Research Service, United States Department of Agriculture (ARS), the average per caput consumption of beef is 157 g per day in the US. Calculations show that 157 g of beef from an animal implanted 61 days before slaughter with a combined implant containing 20 mg oestradiol-17β + 200 mg progesterone or testosterone will contain 3.43 ng oestrogen and 19.5 ng progesterone or 16 ng testosterone.

Table, which provides data on normal levels of these hormones in certain dairy foods, shows that some foods represent hormone sources vastly richer than meat from hormonetreated animals. Based on these values, and averages for consumption of various foods, the relative contribution of meat from hormone-treated animals to the total consumption of hormones has been calculated on the assumption of proper use of the hormones.

It is clear that in most cases the contribution from meat of treated animals is insignificant when hormones have been properly used, and must be considered to be biologically without impact. This becomes even more evident when seen in relation to normal endogenous hormone production in man, as illustrated in Table. It will be seen that even for oestrogens, the hormones considered the greatest risk, the maximal contribution from meat (assuming proper use of the hormones) is less than 0.01% in the prepubertal boy who represents the lowest endogenous oestrogen production.

Table. Hormones in Certain Dairy Foods

	Oestrogens(pg/ml)	Progesterone(ng/ml)
Milk, from non-pregnant	80	9.5 cows
Milk, from pregnant cows	126	
Cream		73
Butter		133
Source: 103.		

Thus far the discussion has been limited to the natural hormones. For synthetic substances the situation may be different. But again, considering the very low residue levels found when hormones have been properly used, the question may be raised whether the risk to the consumers is being grossly overestimated.

Table. Relative Ccontribution of Meat from Hormone-treated Steers to Total Hormone Intake Via Food (per cent)

	Oestrogens	Progesterone
Child under 1 year	0.22	0.014
Child 6 to 8 years	1.56	0.1
Adult male	7.69	0.5

Source: Condensed from 103.

Table. Contribution of Hormones from Hormone-treated Steers Relative to Total Daily Hormone Production in man[1] (per cent)

	Oestrogens	Progesterone	Testosterone
Prepuberal girls	0.00636	0.00078	0.005
Prepuberal boys	0.00826	0.00130	0.00244
Women			
Follicular phase	0.00018	0.00047	
Luteal phase	0.00007	0.01	0.0004
Men	0.00025	0.00048	0.00003

ALTERNATIVES TO THE USE OF HORMONES

Growth rates are influenced by many factors, especially genetic constitution and feeding. Over time, selection as well as improvements in management systems, feed composition and feeding programmes have contributed much to increasing productivity in meat as well as milk. Although it is difficult to evaluate the exact relative contributions of these factors, the overall improvements have been dramatic. An example is the increase in milk yield per head in US dairy cattle. In the period 1944–1975, the number of dairy cows decreased by 33%, while the average yield per cow increased by 60%.

These gains represented a saving of about 23 billion kg of total digestible nitrogen per year, the volume of milk produced remaining relatively constant. The saving is equivalent to about 1.1 billion bushels of maize. Data illustrating progress in beef production over the years are scarce, but increases in productivity similar to those for milk production are unlikely.

In addition to the use of hormones, many avenues are still open for increasing productivity in meat and milk production, including breeding programmes, regulation of rumen fermentation, optimalization of the balance between the indirect and direct feeding of the ruminant organism proper, and disease control.

Breeding programmes

Systematic selection of high-quality sires, combined with an increase in the number of offspring from high-yielding females through embryo transfer, may bring about further improvements in beef and milk production. In many countries, development along these lines has hardly begun.

However, the establishment of effective breeding associations and the strict organization of programme planning and execution are prerequisites for realizing the potentials in this sector.

Regulation of rumen fermentation

The microbial systems in the rumen are extremely complex, and the balance between the various strains of bacteria is susceptible to changes brought about by many factors. The recent introduction of substances such as monensin offers great promise in altering the fermentation pattern to the benefit of productivity by increasing FCE. Since the very extensive breakdown of carbohydrates and protein represents loss of much energy, research is currently being conducted in many laboratories in order to find new methods of increasing FCE.

Optimalization

To a large extent, feeding a ruminant means feeding the rumen microbes which then themselves serve as feed for the organism proper. This is indirect feeding, expensive in energy. On the other hand, the ruminant possesses, in the postruminal part of its digestive tract, all the enzymes necessary for utilizing all types of nutrients except cellulose. The rumen microbes are necessary for the utilization of cellulose, which globally represents an enormous source of energy. However, it is possible to sustain an adequate microbial population in the rumen even when ruminal breakdown of part of the easily digestible nutrients is prevented.

Enabling nutrients to bypass the rumen will increase the utilization of feed for production, and also create a more adequate supply of amino acids. Increased rumen bypass of nutrients can currently be brought about by several means, including formaldehyde and heat treatment of protein-rich feeds. A third method, aiming more at specific substances that may be rate-limiting for production (*e.g.* certain amino acids), or of significance in treatment of diseases, is protection against rumen degradation by such means as incorporation into the ration of long-chain fatty acid mixtures in the form of small pellets. In the future, new methods of increasing rumen bypass will undoubtedly contribute significantly to increased productivity of ruminants.

Disease control

Whatever management system is adopted, effective disease control is essential for productivity. In many areas of the world, infectious and parasitic diseases inflict heavy losses on animal production. A recent study has disclosed nearly a one-to-one relationship between investment in agricultural research and annual productivity of edible protein in ruminants. An increase of about 45% in scientist/man years and a corresponding increase in funding for research and development is considered sufficient to raise productivity in this sector by 50%.

Investment in disease control is an important aspect of this work. Annual world mortality losses from disease exceed 50 million cattle and buffalo, and 100 million sheep and goats. Non-lethal diseases are believed to lead to an equivalent reduction in production. Thus, investment in disease control holds great promise for future augmentation of animal protein production.

In these perspectives, the significance of hormones in animal production may seem marginal, leading to the question of what priority to give to the various efforts to increase productivity and production. In the global context it is, however, at least at present, impossible to adopt one approach to the exclusion of others. As long as preparations exist that combine positive effects on yield and feed utilization with low or non-existing risk to the consumer, there will be a market for them.

What is more, the use of hormonally-active substances in the future may not be limited to those currently available. Common to the present compounds, natural or synthetic, is that they are degraded in the body only to a limited extent. An entirely different situation exists for proteid hormones, which are broken down completely to amino acids, leaving no residues whatever. An example is the growth hormone which not only stimulates growth but also milk secretion, even in high-yielding cows.

This anabolic hormone is currently available only in small quantities for research. However, a recent breakthrough in the use of recombinant DNA technique has made large-scale microbial production of species-specific peptide hormones a realistic possibility. Combined with the development of miniaturized automatic delivery systems for subcutaneous use, a new era may be visualized as regards the use of hormones in animal production.

EXCESSIVE USE OF HORMONES IN ANIMAL HUSBANDRY

Hormones which are intended for occasional therapeutic use are injected into young animals such as calves to fatten them faster. Like athletes taking steroids, this practice results in "fast flesh", but the side effects are unclear. Some single out specific hormone levels and combinations as dangerous to human consumers, while others insist that any hormone additive is potentially dangerous.

Practically, it is an open question whether beef raised with no hormones at all can compete with the quality and market price of beef raised with some hormones, although consumers are now reacting against the flabby, tasteless quality of "hormone-plumped up" meat. The practice of using hormones to improve the growth rates of animals and reduce the fat content of the meat became fashionable during the 1970s, when farmers found they could cut costs as well as satisfy consumer demand for leaner meat.

Bovine stromatotrophin (BST) is a synthetic, genetically engineered substance based on a naturally occurring bovine growth hormone and injected into dairy cows to increase production of milk. The hormone has been widely criticized for its detrimental effects on animal welfare and could pose a health hazard to those who drink the milk.

Whilst widely used in the USA, Canadian and European resistance to the use of the hormone hinges in part on the failure to agree on an international standard for use. The sale and marketing of BST has been banned in the European Union since 1990. The moratorium expired at the end of 1999 but a renewal was requested from the EU Council of Ministers Milk from cows treated with recombinant bovine growth hormone (rBGH in the US, rBST in Europe) has significantly elevated levels of IGF-I (insulin-like growth factor-I), a growth factor responsible for a wide variety of biological functions. In addition, the IGF-I in the milk of rBGH-treated cows is potentially more bioactive than the naturally occurring form, and this bioactivity may be increased further by pasteurization.

Children may be particularly vulnerable to adverse effects, because of their rapid growth rate and their more permeable intestinal tract. In addition, other segments of society may also be at an increased risk, such as those with celiac disease, Crohn's disease, autism, cirrhosis, cow's milk allergy and people taking certain medications The hormones 17beta-oestradiol, progesterone, testosterone, zeranol, trenbolone and melengestrol acetate (MGA) promote growth in cattle. According to a 1999 European Commission report, these hormones have adverse developmental, neurobiological, genotoxic and carcinogenic effects on people who consume them, either via the parent compound, or via metabolites. 17beta-oestradiol both initiates and promotes cancerous tumours. Prepubertal children are at the greatest risk. Growth hormones have been banned in the European Union since 1988. However, many banned substances continue to enter from South America and eastern Europe. Organized crime controls the traffic throughout Europe with production sources through chemical companies and the pharmaceutical industry. Hormones banned in the EEC/EU are held by some farmers to be necessary to the production of high-quality beef at competitive prices. At a time of beef surplus, the EEC/EU banned all hormones, reputedly to discourage farmers from producing beef. There is not a fundamental conviction among farmers in the EEC/EU that the total ban is necessary or economically viable, and this makes the ban difficult to enforce, however needed it may be for public health reasons. In 1999, the USA was using threat of prohibitive tariff protection on European imports if Europe does not drop its restrictions on hormone treated beef exports from the USA. It called the arguments against use of hormones "phony science" Stilbene oestrogens used in animal production as anabolic agents are orally active, persist in food, pose environmental problems because of their low biodegradability. Diethyl stilbesterol (DES) is a known carcinogen. On reaching adolescence, the children of mothers who had taken DES in pregnancy were developing cancer, the girls in the vaginal tract and the boys in the prostate, bladder or testicles. Sometimes the cancer showed as early as seven, sometimes as late as the mid-twenties. Abnormalities showed in 1978 in Milan in children who were fed veal which had been treated with DES. Knowledge of the toxicology of many other used substances is still rudimentary

- Some synthetic hormones, such as diethyl stilbenes, are notorious cancer causing agents
- Under European law, substances can be administered to livestock only for veterinary purposes or on grounds of hygiene, unless they are not harmful to their health or welfare. This does not seem to be the case as regards BST
- There is no documented case where use of antibiotics in livestock has been proved to cause treatment failure in humans. Without growth promoters Europe would need more land to grow animal feed, be swamped by an extra 7 million cubic metres of waste manure each year, and face additional emissions of up a third of methane, nitrogen, phosphorus and ammonia, because growth promoters make animals' dietary systems more efficient

LEVELS OF ENDOGENOUS HORMONES IN BODY FLUIDS AND TISSUES

Any discussion of possible health hazards connected with the use of hormones in animal production must take into account the normal occurrence of hormones and their metabolites in body fluids and tissues, and the fact that the levels of these hormones vary greatly, according to the physiological state of the animal.

Table. Concentrations of Endogenous Hormones in Edible Tissues of farm Animals

Animal/tissues	Oestrone pg/ml	Oestradiol-17αpg/ml	Testosterone pg/g	Progesterone pg/
Veal calf				
muscle		< 100	70	
liver		< 100	47	
kidney		< 100	685	
fat		< 100	340	6
Bull				
muscle			335	
liver			749	
kidney			2 783	
fat			10 950	
Heifer				
muscle		12–13	92	
liver	20–40	38–71	193	16
kidney		40–71	595	
fat		6	250	
Cow, pregnant		370–860		
muscle		2 500–5		336
fat	3 870	500		
Steers				
muscle	6	14		
liver	20	14		
fat	23	10		
Wheat germ oil	4 000 pg/g DES equivalent			
Soy-bean oil	2 000 000 pg/g DES equivalent			

Thus, oestrogen levels in the blood of female farm animals may vary from a few pg up to 5–6 000 pg per ml plasma. As to males, the plasma of stallions and entire male pigs contains high levels of oestrogens, although mainly in the conjugated form. Milk also contains oestrogens in very high concentrations in the first drawings after parturition; in non-pregnant animals, levels in the range of 80–100 pg/ml have been reported. More recently, reliable data have also become available concerning concentrations in edible tissues; some of these are presented in Table. For the sake of comparison, levels of oestrogen activity normally present in products of plant origin widely used in human nutrition are included.

METABOLISM, ROUTES AND RATES OF ELIMINATION

The general patterns of metabolism and elimination of endogenous hormones in farm animals have been outlined. In ruminants, testosterone and oestradiol-17β are rapidly converted to their epimers, biologically much less active, epitestosterone and oestradiol-17α.

Progesterone is partially converted to androgens before excretion. In the pig, epimerization of testosterone and oestradiol-17β does not appear to take place to a significant degree. The faecal route of elimination dominates in ruminants, while in the pig urinary excretion is more important.

Progesterone

After repeated injections of progesterone to cows and steers over 2 to 3 weeks followed by ^{14}C-progesterone for 2 to 5 days, the animals were slaughtered 2 to 3 hours after the last injections. Activity levels were 2 to 7 times higher in the fat, 3 times higher in the kidneys, and 13 times higher in the liver than in the muscle. Excretion of radioactivity amounted to 50% and 12% in faeces and 2.0% and 1.2% in urine in cows and steers respectively. About 50% of the activity in muscle and milk was associated with unchanged progesterone, most of the remaining activity being associated with a mono-hydroxy compound. Cooking or frozen storage did not affect the nature or quantity of metabolites.

Oestradiol-17β

Following daily injections of 1 mg oestradiol-17βor its benzoate to heifers and steers for 11 days, followed by the ^{14}C-compounds on days 12, 13 and 14, the animals were slaughtered 3 hours after the last injections, when residual levels were maximal. In muscle extracts, oestradiol-17β represented the major fraction of extracted activity (38 to 71%), followed by oestrone (17 to 45%). Levels in muscle were 161 to 225 pg/g and 40 to 86 pg/g for oestradiol-17β and oestrone respectively. In fat the levels were 3 to 5 times higher.

The authors conclude that residual levels are extremely low when these hormones are administered as growth stimulants to growing/finishing cattle.

Glucosides of the 17β- and the 17α- epimers, and the glucoronide of the 17α-epimer are the major metabolites in cattle. When oestradiol-17β was administered orally to swine, plasma concentrations were very high 7 min after administration. Oestradiol was completely conjugated during absorption and its first passage through the liver. Some conversion to oestrone took place.

DES

The metabolism of DES in food-producing animals has been reviewed recently. The substance seems to be eliminated to a large extent in unaltered form. After oral administration of ^{14}C-DES to beef cattle, 99.5% of the radioactivity was excreted within 5 days after withdrawal. In liver extracts, radioactivity associated with DES-conjugate and free DES was found to be 75% and 25% respectively.

Higher than background levels of activity were observed after withdrawal in kidney, liver, bile and urine/faeces for up to 5, 7, 9 and 11 to 12 days respectively. The fate of 24-mg DES implants containing ^{14}C-DES and implanted in the dewlap of calves was studied over 98 days. Free radioactivity was almost completely associated with unchanged DES. At the time of slaughter, levels were less than 0.1 ppb in muscle and fat, and 1 to 1.5 ppb in liver and kidney. In a study in steers implanted with ^{14}C-DES, on day 120 after implantation radioactivity in muscle was not distinguishable from background. It was above background in spleen, lung, adrenal glands and kidney, but less than levels corresponding to 0.5 ppb.

In a similar study on steers, 120 days after implantation, levels in liver, kidney, lungs and salivary glands were in the range of 0.07 to 0.13 ppb of DES equivalent. In a recent study of DES metabolism in rhesus monkeys and chimpanzees, most of the substance was excreted with the urine. Extracts in the organic and aqueous phase mostly contained unchanged DES in the free and conjugated form respectively. Current evidence indicates that the oxidative metabolism of DES leads to at least three compounds that may have cytotoxic or mutagenic activity, but these have not been identified as DES metabolites in ruminants, but in the mouse.

Zeranol

Using a gas chromatographic method with a sensitivity limit of 20 ppb, no residues of zeranol could be detected in edible tissue from cattle slaughtered 65 days following implantation of 36 mg, or from lambs 40 days following implantation of 12 mg. In another study, tritiated zeranol was implanted in cattle as part of 36-mg doses. Skeletal muscle obtained 10, 30 and 50 days following implantation contained no detectable residual activity. This confirms previous results based on the use of ^{14}C-labelled zeranol.

Trenbolone acetate (TBA)

Trenbolone is a 17β-OH steroid esterified in the 17 position with acetic acid. Upon release in the organism the ester is rapidly hydrolysed to the free compound TB-17β-OH and acetate. In cattle the 17β-OH compound is rapidly transformed to its 17α-OH epimer, in the same manner as oestradiol-17β in this species.

The 17β epimer possesses only about 5% of the biological activity of the 17β epimer. Another metabolite of TBA in cattle is the 17-keto compound, analogous to oestrone; quantitatively it appears to be of very little importance. Following intravenous injection of TBA, levels of TB-17β-OH and TB-17α-OH of 0.05 and 0.005, 0.10 and 1.0, 0 and 191 ppb have been recorded for muscle, liver and bile respectively. Other metabolites occurred in extremely small quantities in cattle. Similar findings have been reported in studies based on the use of implants. The major route of excretion is by faeces. Metabolism studies of TBA thus clearly show that the substance is rapidly subjected to biological inactivation in cattle, mainly by epimerization of the free steroid to the 17α-compound, and that the major route of excretion is via the bile.

HORMONE PREPARATIONS

Hormones of endogenous origin

These comprise the "classical" steroid sex hormones, oestradiol-17â, testosterone and progesterone. The two former are used either in the free form or as esters, mainly those of propionic or benzoic acid. Esterification generally causes prolongation of the half-life of the compounds in the body by 40 to 50%. The natural hormones having low bioavailability when administered orally, owing to rapid conjugation and metabolic transformation in the liver, they are therefore administered by subcutaneous implantation.

Hormones of exogenous origin

Of the oestrogens, the stilbene derivatives diethylstilboestrol (DES) and hexoestrol possess high biological activity and have been used most widely. They are active orally as well as by implantation. Other orally active oestrogens include ethynyl-oestradiol, a more slowly metabolized derivative of the true hormone, with higher activity. An oestrogen with an entirely different structure is zeranol, a derivative of a resorcylic acid lactone occurring in the fungus

Giberella zeae. The synthetic androgens comprise a large number of substances, most of which are steroids. Of these, trenbolone acetate (TBA) possesses strong anabolic properties and has received much attention during recent years, used alone or in combination with an oestrogen. Another anabolic steroid is methyl-testosterone. Of synthetic gestagens, only one will be mentioned here: melengestrol acetate, which stimulates growth in heifers but not in steers, and which can also be used for the suppression of oestrus. Numerous other gestagens also exist, but at present few other than progesterone and melengestrol acetate are used to stimulate growth.

In addition to these substances, numerous others exist, and some of them are used more or less frequently in clinical veterinary medicine. However, clinical applications of hormones are not considered to be of consequence to the consumer, since such treatment is much less frequent than the use of hormones to promote growth.

Hormone preparationsin current use as growth stimulants are listed in Table, which also shows modes of application, dosages, etc. It will be noted that almost all preparations currently in use are based on implantation, the site usually being the base of the ear, or less frequently, the dewlap.

EFFECTS OF HORMONES

Veal calves

In veal calves, hormone treatment may begin at a body weight of about 65 kg, the animals being slaughtered at about 170 kg. Implants of 20 mg oestradiol-17β + 200 mg progesterone in males and 20 mg oestradiol-17β + 200 mg testosterone in females resulted in a 20% increase in daily gain and 21% higher nitrogen retention in the period studied . In other studies, improvements of 10 to 12% in gain and 10% in FCE have been reported.

Nitrogen retention is about 70% in the very young veal calf, but decreases gradually to below 40% at the age of about 15 weeks. For ages of 10 to 15 weeks, the average conversion of feed protein to body protein is about 40%; this rate can be increased to about 60% by hormone treatment.

The effective preparations were DES, oestradiol-17β, and the combination of TBA + oestradiol-17β. More recently, positive effects have been reported for zeranol alone (36 mg) and for zeranol (36 mg) + TBA (140 mg), with increases in nitrogen retention of the same order as for DES and E_2 + TBA. When zeranol + TBA was implanted at the age of 56 days, the growth rate up to day 106 increased by 18% .

Steers

The most extensive studies of the effects of hormones on growth and FCE have been carried out on steers, under strictly controlled conditions as well as in the field. Since 1975, most studies have involved implants of oestrogens alone, androgens alone, or combined oestrogen/androgen preparations, although many trials have also been based on oestrogen/progesterone combinations during recent years. Oestrogen implants have included DES, hexoestrol, oestradiol-17â and zeranol. DES implants have, as in previous studies, resulted in an increase of about 12% in gain and in improvement in FCE of the order of 10%. Hexoestrol implants, usually in doses of 30 to 60 mg, have been shown in numerous experiments to lead to considerable improvement in growth rate and FCE. In 19 trials carried out over the years on experimental husbandry farms in the UK, the overall average increase in gain produced by 45 or 60 mg hexoestrol implants was 0.16 kg a day, and in only 2 of the trials was it less than half that figure. Oestradiol - 17â implants alone (30 mg) have resulted in a 24% increase in gain and a 13% improvement in FCE.

Zeranol implants, usually at 36 mg, have consistently improved gain as well as FCE. In a series of 21 UK trials over several years, the average response to zeranol implants alone was an increase in daily gain of 0.15 kg. Only in one trial was there no response . Similar results have been obtained in Ireland. Positive effects on gain in steers have been observed under a variety of experimental conditions, under controlled feeding, on ad lib feeding of standardized rations, and on pasture.

TBA implants administered alone at a dose of 300 mg have also had positive effects on growth, even if combination with an oestrogen has yielded better responses (vide infra). In a series of 8 trials in the UK, the average additional daily gain amounted to 0.09 kg, with considerable variation among trials . Similar results have been reported from Ireland. Combined preparations. A number of trials have been carried out with implants containing two hormones.

The combination of an oestrogen with an anabolic steroid, or with progesterone, has met with the greatest responses. Synovex-S has consistently increased gain as well as FCE, with responses averaging about 20% and 17% respectively. Hexoestrol + TBA (usually 30 or 45 mg hexoestrol + 300 mg TBA) has resulted in marked increases in gain, of the order of 30% and in FCE of the order of 20%. Oestradiol-17? + TBA (20/140 mg) has given similar results, as has Zeranol + TBA (36/300 mg), also recently tested.

Hormone preparations have also been tested in combination with substances such as monensin, which increase FCE by promoting propionic acid formation in the rumen. Results have varied from no effect to marked additional gain. Reimplantation, tested under various forms of management with varying results, has not gained general acceptance. Lamming has stated that "repeat implantation of hormone is not likely to produce the benefits obtained from its initial use, since a second dramatic change in the endocrine balance of the animal is not likely to occur. In addition, double implantation increases the possibility of exceeding the optimum dose rate and the chance of deleterious side effects occurring." The evidence for highly significant positive effects on the growth rate and FCE of steers is thus beyond dispute, the most marked effects being provoked by implants combining an oestrogen with an androgen of high anabolic activity.

Bulls

Since the entire male animal produces its own anabolic androgen, testosterone, an effect of additional hormones similar to that for steers is not to be expected. The number of trials with bulls is also limited. Positive effects on gain have been reported using DES alone and combined oestrogen/TBA implants ; in other studies, no effect on gain has been recorded, while a certain increase in the deposition of fat in the carcass has been observed.

Undesirable side effects in treated animals

Reported side effects of hormone treatment for growth stimulation are few and generally concern the use of oestrogens in steers. Changes in body conformation such as feminization and raised tail-heads were described as early as 1958. Similarly, bulling has occurred with increased frequency, although in most animals it is limited to the first few days after implantation.

However, it has been reported from Kansas that 2.2% of all steers fed in pens have to be removed, at an estimated loss of $23 per head. In a study of the effect of reimplantation of oestrogens in steers, all animals were given a 30 mg DES implant at a live weight of 260 kg, and then reimplanted 91 days later, with either 30 mg DES or Synovex S. Following the second implant, the frequency of the steer-buller syndrome was 1.65% for the DES-DES group, and 3.36% for the DES-Synovex S group. The economic advantage of using DES + DES was estimated at $1.15 per head. The steer-buller syndrome is a special problem in feedlots.

Bibliography

Arseniev, V.A.: *Atlas of Marine Mammals,* Neptune City: T. F. H. Publishing, Inc., 1986.

Baker, Steve: *The Postmodern Animal.* London: Reaktion Books, 2000.

Bekoff, Marc: *Strolling with Our Kin: Speaking For and Respecting Voiceless Animals.* New York: Lantern Books, 2000.

Blomquist, G. J.: *Pheromone Biochemistry.* Academic Press, New York, 1987.

Bower, B.: *Fossils may Clarify Mammal Evolution,* Science News, 1984.

Broach, J.R.: *The Molecular Biology of the Yeast,* Cold Spring Harbor, Cold Spring Harbor Laboratory, 1981.

Brock Juliet : *Horse power: a history of the horse and donkey in human societies,* National history Museum publications, London 1992.

Brock, J. : *A Natural History of Domesticated Animals,* Cambridge Univ. Pr., New York, 1999.

Bronson, F. H.: *Mammalian Reproductive Biology,* Univ. Chicago Pr., Chicago, 1990.

Brown, L.: *Cruelty to Animals: The Moral Debt.* London: MacMillan, 1988.

Burny, A : *Biotechnology in Animal Husbandry,* Springer Pub, 2008.

Buss, D. M. : *The handbook of evolutionary psychology.* Hoboken, NJ: John Wiley, 2005.

Charles L. : *The Cycles of Plant and Animal Nutrition,* Scientific American Books, San Francisco 1976.

Chhazllani, V K: *Dairy Chemistry and Animal Nutrition,* Manglam Pub, Delhi, 2008.

Clymer, R. : *Nature's Healing Agents,* PA, U.S.A: Dorrance Co., 1963.

Crawford A. : *Experiments and Observations on Animal Heat,* London: Printed for J. Johnson; 1788.

Daniel, J.C. Jr. : *Methods in Mammalian Reproduction,* Orlando, FL, Academic Press, 1978.

Darlington, C. D.: *The Elements of Genetics*: London, Allen and Unwin, 2007.

Darwin, Charles: *On the Origin of Species* (1st ed.). London: John Murray. 1859.

Dawkins, R.: *River Out of Eden. A Darwinian View of Life,* Basic Books, Harper Collins. New York, 1995.

Degen, A. A. : *Ecophysiology of Small Desert Mammals,* Springer, New York, 1997.

Devyani Khemka: *Animal Physiology,* Dominant, Delhi, 2003.

Elliott, Daphne C.: *Biochemistry and Molecular Biology,* Oxford University Press, Delhi, 2005.

Else PL. : *Mechanisms Underlying the Cost of Living in Animals,* Annu Rev Physiol. 2000.

Ensminger, M.E. : *Dairy Cattle Science,* The Interstate Printers & Publishers, Inc., Danville, 1980.

Frederick B.: *Genetics for Dog Breeders,* Freeman & Company, 1979.

Glut, D.F.: *The Frankenstein Legend: A Tribute to Mary Shelley and Boris Karloff.* Metachen, New Jersey: The Scarecrow Press, 1973.

Gordon, G. A. : *Animals Physiology,* Harper and Row, New York, 1989.

Greene, H. W.: *Mode of Reproduction in Lizards and Snakes of the Gomez Farias Region,* Tamaulipas, Mexico. Copeia, 1970.

Hacker, J.B. : *Nutritional Limits to Animal Production from Pasture,* Farnham Royal: CAB, 1981.

Hagedorn, A.L.: *Animal Breeding,* Crosby Lockwood, 1950.

Hames, B. D.: *Genes and Embryos*: New York, Oxford University Press, 1989.

John E.: *The Animal Radiations,* The University of Chicago Press, 1981.

Joysey, K. A. : *Development, Function and Evolution of Animal Teeth,* Academic Pr., New York, 1978.

Lata Bhattacharya: *Animal Biochemistry,* Discovery, Delhi, 2010.

Lewis, Ricki: *Human Genetics: Concepts and Applications,* Dubuque, IA: McGraw-Hill Higher Education, 2001.

McKelvey, J. J. *Chemical Control of Insect Behavior.* Wiley-Interscience, New York, 1977.

Mindell, E. : *Mindell's Vitamin Bible,* New York, U.S.A: Warner Books, 1980.

Muybridge, E. : *Muybridge's Complete Human and Animal Locomotion,* Dover Publ., New York, 1979.

Orians G, Heller C. :*Life: The Science of Biology,* Massachusetts: Sinauer Associates, 2002.

Plomin, Robert: *Behavioral Genetics,* San Francisco: W. H. Freeman and Company, 2000.

Richard M.: *Understanding Animal Breeding,* Prentice-Hall, 2000.

Richards, R. J. & Ruse, M. : *Darwinian heresies.* Cambridge, UK: Cambridge University Press, 2004.

Ridley, M.: *Genome: The Autobiography of a Species,* New York: HarperCollins, 2005.

Stwertka, Eve : *Genetic Engineering,* Franklin Watts, New York, 1982.

Thornhill, R. : *The Evolution of Insect Mating Systems.* Harvard University Press, Cambridge, Massachusetts, 1983.

Wilson, Edward O. *On Human Nature.* Cambridge, MA: Harvard University Press, 1978.

Wilson, P. : *Genes, Mind and Culture: The Coevolutionary Process.* Cambridge, MA: Harvard University Press, 1981.

Index

❑❑❑